EurographicSeminars

Tutorials and Perspectives in Computer Graphics

Edited by W. T. Hewitt, R. Gnatz, and D. A. Duce

P. J. W. ten Hagen P. J. Veerkamp (Eds.)

Intelligent CAD Systems III

Practical Experience and Evaluation

With 116 Figures and 5 Tables

Springer-Verlag
Berlin Heidelberg New York
London Paris Tokyo
Hong Kong Barcelona
Budapest

EurographicSeminars

Edited by W. T. Hewitt, R. Gnatz, and D. A. Duce
for EUROGRAPHICS -
The European Association for Computer Graphics
P. O. Box 16, CH-1288 Aire-la-Ville, Switzerland

Volume Editors

Paul J. W. ten Hagen
Paul J. Veerkamp
Interactive Systems
Centre for Mathematics and Computer Science
P. O. Box 4079
NL-1009 AB Amsterdam, The Netherlands

ISBN-13: 978-3-642-84394-5 e-ISBN-13: 978-3-642-84392-1
DOI: 10.1007/978-3-642-84392-1

Library of Congress Cataloging-in-Publication Data
Intelligent CAD systems III: practical experience and evaluation / P. J. W. ten Hagen, P. J. Veerkamp (Eds.) p. cm. - (EurographicSeminars) "Papers presented at the Third Eurographics Workshop on 'Intelligent CAD Systems' which was held April 3-7, 1989 at Hotel Opduin on the island Texel in the Netherlands" - Pref. Includes bibliographical references.
ISBN-13: 978-3-642-84394-5

1. Computer-aided design - Congresses. 2. Engineering design - Data processing - Congresses. 3. Artificial intelligence - Congresses. I. Hagen, P. J. W. ten. II. Veerkamp, P. J. (Paul J.), 1959-. III. Eurographics Workshop on "Intelligent CAD Systems" (3rd: 1989: Texel, Netherlands) IV. Title: Intelligent CAD systems 3. V. Series: Eurographic Seminars.
TA174.I468 1991 620'.0042'0285 - dc20 91-10006 CIP

Softcover reprint of the hardcover 1st edition 1991

45/3140-5 4 3 2 1 0 - Printed on acid-free paper

Preface

This book contains a selection of revised versions of papers presented at the Third Eurographics Workshop on Intelligent CAD Systems, which was held at Hotel Opduin on the island of Texel in The Netherlands, April 3–7, 1989. The workshop theme was *Practical Experience and Evaluation*. It included five paper presentation sessions, each followed by a discussion. The workshop closed with a general discussion. The book is therefore divided into five parts: *design process, system architecture, languages, geometric reasoning,* and *user interface*. A report on the discussion session, written by the session's moderator, concludes each part. These reports are not intended to be exact records of the discussion, but rather the moderators' summary of their contents.

The aim of the workshop was to share the experience the participants gained by developing *intelligent CAD* (Computer Aided Design) systems, and to evaluate the developed systems to determine which features were still lacking. The workshop was organized as the last one in a series of three workshops under the same title. The first workshop focused on theoretical and methodological aspects, resulting in a sound theoretical basis for intelligent CAD systems. Implementational issues were discussed at the second workshop, paying attention to systems developed with reference to this basis. The experience and evaluation showed a dual outcome. Firstly, it resulted in the development of a new generation of intelligent CAD systems. Secondly, it led us to the development of new theories for intelligent CAD.

The first paper of the opening session on the *design process* took us deep into logical abstraction, focusing on formal expressions of requirements and inference processes that can generate solutions. The second paper referred to protocol studies and brought us face-to-face with users (designers). The last paper focused attention on a specific technique for integrating multiple contributions to designs of single products. The three papers cover a broad range of interest, from formalisms in logic, through design behaviors in persons, to an elaboration of technical implementation issues in computer systems. Logic offers a neat clarity which is appealing, and applications in user domains perhaps inevitably entail confusion. We need to establish robust and general logical treatments that are meaningful to designers and which can survive complexity and scale of applications.

The second session on *system architecture* started with a paper emphasizing the problems of persistence associated with large knowledge bases. The second paper presents ideas on the use of functions to match input and output of processes and the application of these ideas in conceptual design and requirements analysis. In the third paper, an argument and approach for involving and using a design history mechanism is presented. The common thread between these papers might be that their focus of attention is now closer to the ideal system discussed at the first two workshops.

The third session on *languages* started with a paper emphasizing the representation of design processes in ADDL (Artifact and Design Description Language). The use of formal logic in ADDL and the concept of scenarios seemed to draw attention. The second paper describes a representation language (GRAFLOG) based on formal logic for capturing CAD knowledge. It demonstrates the use of this language in dealing with the semantic contents of the dialogue between a system and a designer. The third paper describes an experience with development and use of IDDL (Integrated Data Description Language) at the University of Tokyo, and discusses examples of design knowledge representation in this language. The IDDL work at the University of Tokyo is now a parallel effort with the ongoing project at CWI (ADDL). It is clear that the two implementations are different, and it seems there are even some conceptual differences between the two projects. A common thread is that each of the three papers treats a representation language for intelligent CAD based on a combination of logic and objects.

The fourth session on *geometric reasoning* started with a paper which studies the geometric reasoning module as part of a complete system for geometric analysis. A user requests some predefined tasks to be performed, where each task is translated to a combination of geometric modeling, reasoning and motion planning activities. The second paper uses objects and constraints in 2D-space planning. It concentrates on search strategies over the state space of partial solutions. The user interface allows for close observation of the search process and for providing directions during the search. The third paper proposes a constraint management system to experiment with constraint-based design for mechanical engineering applications. The system supports geometric objects with mutual relations represented as constraints. For each constraint, operations are defined on the objects to satisfy it and to retain it when additional operations and/or constraints are imposed.

The last session on the *user interface* started with a paper advocating an instrumentalist approach to computer systems. It particularly stresses the importance for a designer to express his[1] thoughts in both words and pictures. A designer must therefore be given a tool which allows him to construct expressions that give him control over the behavior of the system. The second paper specifies a data-structure for storing a design part together with its geometrical and technological attributes, in order to achieve the possibility of interaction between CAD and NC systems. It also offers a tool for the graphical representation of knowledge, based on an object-oriented approach. The third paper focuses on a dialogue language for human computer interaction. The author argues that a system is entitled to perform certain tasks autonomously, provided that the designer is always informed about what the system is doing. A monitor component, with models both of the user and the system, guides the communication between them. The discussion focuses on the autonomy of an intelligent CAD system, which should never ''get in the way'' of designers. The intelligence of a system, therefore, lies in its understanding the designer's demands and acting adequately upon his commands and *not* in performing certain design tasks autonomously.

We would like to thank all authors for their contributions, the workshop participants who made the workshop interesting and took active part in the discussions, and the organisers who

[1] Throughout the book the male form refers to both female and male.

made the workshop possible. The work of the programme committee members is very much appreciated in ensuring the high quality of the programme and in selecting diverse papers. The Centre for Mathematics and Computer Science (Centrum voor Wiskunde en Informatica, CWI), the official organiser of the workshop, allowed us to use its modern computer equipment and high quality text processing facilities to generate this output. The workshop itself was partially funded by NFI (Nationale Faciliteit Informatica), a Dutch organisation for promoting research in computer science, and CWI. Last but not least we would like to express our gratitude to Marja Hegt who was the workshop secretary.

Amsterdam, October 1990

Paul ten Hagen
Paul Veerkamp

Programme Committee of the Eurographics Workshops on Intelligent CAD Systems

(As of December 1989)

Table of Contents

Geometric Reasoning

User Interface

Design Process

1

A Logical Framework for Design Processes

Jan Treur

Expert Systems Section
Programming Research Group, University of Amsterdam
Kruislaan 409, 1089 SJ Amsterdam, The Netherlands

Abstract: Designing often involves rather complex processes of reasoning. For building knowledge-based systems for designing, a detailed logical analysis of these processes is inevitable. This paper aims to present a logical framework for describing the steps in a design process as logical inferences. To represent the various kinds of reasoning involved, this framework uses reasoning about viewpoints and reasoning about partial models. These are particular kinds of meta-level reasoning. The logical framework, which is presented in this paper, offers a basis for a formal specification of knowledge-based systems for design.
Keywords: Design, Logical Analysis, Knowledge-Based Systems, Complex Reasoning, Meta-Level Reasoning, Formal Specification

1. Introduction

In general, a design process starts by establishing the need for a certain object and by expressing a number of properties that the object should possess; we call these properties the *(initial) design requirements*. The aim of a design process is to construct an *explicit description* of the structure of an object satisfying the design requirements. This explicit description needs to be sufficiently detailed to guarantee that the object with this structure actually can be built. However, it may happen that the outcome of a design process is the assertion that there does not exist an object satisfying the requirements (inconsistent design requirements). In this case one may replace some of the requirements by others and restart the design process.

Designing often involves rather complex processes of reasoning. For building knowledge-based systems for designing, a detailed logical analysis of these processes is important and may be inevitable. This kind of analysis has little attention in the literature; some papers even claim that "design is logically impossible" (see [6]). However, it seems that the interest in theoretical and methodological aspects of design processes in general is growing (see [4, 5, 7, 8]). This paper aims to present a *logical framework* for describing the steps in a design process as logical inference steps. To represent the various kinds of reasoning involved, this framework uses reasoning about logical theories (also called reasoning about viewpoints, see [1,2]) and reasoning about *partial models*. These are particular kinds of meta-level reasoning.

For analysing complex reasoning tasks for knowledge-based systems, by Chandrasekaran et al. at Ohio State University a theory of generic tasks is developed; this theory and its applications to design tasks are presented, for instance, in [3]. In literature on this theory of generic tasks no formal description is given. In our work (for instance see [9, 10, 12]) it is discussed how complex reasoning tasks for knowledge systems may be formally modelled by interactions of uniform reasoning tasks. In [9, 10] it has only been sketched how to apply our ideas to complex reasoning tasks in the field of design. In this paper a logical framework is presented which offers a more specific basis for a formal specification of knowledge-based systems for design.

Note that the fact that we use logic for such a formal specification does not imply that we propose Prolog as a language for implementation of these knowledge-based systems. The choice for a suitable implementation language or tool is independent of the chosen specification formalism. This paper is not on the choices to be made for implementations.

2. Incomplete Descriptions of Objects by Partial Models

Constructing an object-description satisfying given design requirements is a process involving a number of steps. Each step fills out more detail of the structure of the object. Also choices may be made and, if a choice turns out wrong, backtracking may occur. For describing traces of such a process, descriptions of structures are needed which may be given in different levels of detail. For this purpose we use partial models as known from logic (for instance, see [13], p.68).

2.1. Partial models

We shall discuss partial models for propositional logic and for many-sorted predicate logic. When appropriate we use propositional descriptions. In this case a partial model is an assignment of truth-values true or false to a *subset* of the set of atoms occurring in the logical language; this may be viewed as a partial function from the set A of atoms into the set of truth values:

$$M : A \rightarrow \{\text{true}, \text{false}\}$$

If we assume a certain order, say $A = \{a1, a2, \cdots\}$ then a propositional partial model M may be represented by a sequence of symbols from {true, false, *} or {0, 1, *}. So, for a partial model:

$$< 1, 0, *, 1, 1, 0, *, *, 0, \cdots >$$

a 1 in the n-th position means the n-th atom is true, 0 that it is false and * that it is (as yet) undecided. For finite cases this corresponds to a conjunction of atoms and negations of atoms

$$a_1 \wedge \neg a_2 \wedge a_4 \wedge a_5 \wedge \neg a_6 \wedge \neg a_9 \wedge \cdots$$

A partial model may be viewed as an incomplete description of the structure of an object. It is called *complete* if all atoms have a truth-value assigned to them, i.e. if no * occurs. The *signature* of a partial model is the set of symbols of the language that is considered. For the propositional example above this signature is the set A. In the case of many-sorted predicate logic the signature is the set of all constants, function symbols and predicate symbols in the language under consideration.

2.2. Reducts and expansions

Partial models are closely related to reducts and expansions of (complete) models. Suppose L_2 is a signature extending the signature L_1 and M_2 is a model of signature L_2. The *reduct* M_1 of M_2 is the model of signature L_1 obtained from M_2 by deleting from it the part corresponding to symbols of L_2 which are not in L_1; the model M_2 is called an *expansion* of M_1. If, for instance, M_2 is the complete model

$a_1 \to 1$
$a_2 \to 0$
$a_3 \to 0$
$a_4 \to 1$

of signature $A = \{a_1, a_2, a_3, a_4\}$ then the reduct of M_2 with respect to the restricted signature $A_0 = \{a_1, a_2, a_4\}$ is the model M_1 given by

$a_1 \to 1$
$a_2 \to 0$
$a_4 \to 1$

Formally spoken this simply may be expressed by saying that the restriction of the function M_2 to the set A_0 is M_1. This reduct M_1 is a total function from the part A_0 of the signature A to the set of truth values. Therefore M_1 may be identified with a partial function from A to the set of truth values which is only defined for atoms in A_0. Thus a reduct M_1 of a complete model M_2 of signature A may be identified with a certain kind of partial model of the same signature A. In other words: with respect to the complete signature A the reduct M_1 may be viewed as a partial model; with respect to the reduced signature A_0 it is a complete model. In the propositional case every partial model may be obtained as a reduct of a complete model. In the case of many-sorted predicate logic partial models may occur where some functions or relations are only partially defined in the domain (see [13], §3.2 and §5.4). These are not reducts of complete models. Below in Table 1, a simple example will be given; in this example partial models will be reducts.

Reducts and expansions of partial models may be defined as well. This may be done similar to the definition of reducts and expansions of complete models.

2.3. Refinement relation

Processes of incremental refinement in designing may be described by *incremental construction of partial models* which grow more and more complete. This provides a trace of partial models. Each of these partial models satisfies only a part of the design requirements and its structure is filled out only for a part of the logical language of the final model. This final model is a complete model of a certain signature expanding the signature of the design requirements and satisfying all these design requirements. To describe the steps of refinement in such a trace of partial models we define the *refinement relation* between partial models of arbitrary, possibly different signature: $M_1 \leq M_2$ if all atoms with a truth-assignment in M_1 also have a truth-assignment in M_2

with the same truth values (the truth-assignment of M_2, considered as a partial function, is an extension of the truth-assignment of M_1). The refinement relation is reflexive and transitive. It may happen that for different partial models M_1 and M_2 we both have $M_1 \leq M_2$ and $M_2 \leq M_1$. We call such M_1 and M_2 *equivalent;* this determines an equivalence relation. Examples of equivalent partial models are provided by partial models of signatures S_1 and S_2 where all symbols not in the intersection of the signatures are undefined. This equivalence subsumes the identification between reducts of complete models and partial models as discussed in §2.2.

In design processes incremental refinement may often be viewed as introducing additional structure that cannot be described in the language of the requirements. For instance this is done by describing the object as a structure consisting of a number of independent components with some kind of interaction between them. In this case the refinement is created by introducing an expansion as described above. In the many cases that the design requirements do not explicitly prescribe the structures for such an expansion, the search for a useful expansion may be one of the aims of the design process. It may be a variable in the design process. The example in Table 1 below, illustrates this.

2.4. A simple example

We give a simple, mathematical example; an example in a more realistic context will be given in §5. The example below deals with finding (designing) a positive solution of a quadratic equation. Although this may not be the first issue considered in designing, it is useful to explain a number of our ideas. The design process proceeds as given in Table 1. Here R denotes the model of the real numbers with constants 0 and 1, the operations of addition, subtraction, multiplication and the order relation; the language L is the language for this model extended by the constant x. The design process aims at constructing a model $<R, c>$ for a certain element c of R such that the initial design requirements (of I) are satisfied if the constant x is interpreted as c. Initially c is not given an interpretation (denoted by *); we deal with a partial model. In II to this partial model some additional structure is added: the constant y is added to the language and will get an interpretation d later on (the second *). This expanded partial model is filled in (in III and IV) to a complete model. Here two possibilities are involved (note that the requirements placed between [] are left out of consideration temporary).

Of these two possibilities only one (V) satisfies the requirement $x \geq 0$. This remaining model satisfies all initial design requirements. The solution (in VI) is a reduct of this complete model: deleting y from the language and its interpretation d from the model.

Note that during this design process both the partial models and the design requirements are changed dynamically: in II the requirement $x^2 + 4x - 5 = 0$ is transformed to $y^2 - 9 = 0$. Also some new requirement $x + 2 = y$ is added to describe the connection between the model with x and the model with y. This dynamical behaviour of the design process essentially is the phenomenon we have to describe by our logical framework. Also note that the expansion of the model by the constant y is not determined by the initial requirements. To use just this expansion is a choice which is made in the design process; other choices could be made as well. This is part of the dynamical changing of the requirements.

design requirements and language	models
I. language: L $[x \geq 0]$ $x^2 + 4x - 5 = 0$	 < R, * > * not yet interpreted (constant x)
II. language: L + constant y $[x \geq 0]$ $x + 2 = y$ $y^2 - 9 = 0$	 < R, *, * > * not yet interpreted (x resp y)
III. language: L + constant y $[x \geq 0]$ $[x + 2 = y]$ $y = 3$ or $y = -3$	 < R, *, 3 > or < R, *, –3 > * not yet interpreted (x)
IV. language: L + constant y $[x \geq 0]$ $y = 3 \wedge x = 1$ or $y = -3 \wedge x = -5$	 < R, 1, 3 > or < R, –5, –3 >
V. language: L + constant y $y = 3 \wedge x = 1$	 < R, 1, 3 >
VI. language: L $x = 1$	 < R, 1 >

Table 1. Solving an equation as a design process

By a slight change the example in Table 1 may be transformed into one involving functions. For instance, if the quadratic equation $x^2 + 4x - 1 = 0$ is chosen then the square root function may serve as a structure to express the solution. Another example is solving an inhomogeneous linear differential equation under certain boundary conditions. In this case the solution is some function. A common strategy splits this design problem into two other design problems: (a) determining all solutions of the homogeneous equation corresponding to the given equation without taking the boundary conditions into account, and (b) determining a particular solution of the given inhomogeneous equation, again without taking the boundary conditions into account. As a final solution a combination of these two kinds of solutions is constructed which does satisfy the boundary conditions. This strategy is used more often in designing (hierarchical decomposition; see below).

2.5. World of objects

The world of (realisable) objects will be denoted by W; these are all those formal descriptions which give rise to objects which may be constructed in reality. For instance descriptions concerning objects with very small dimensions, which may exist in a theoretical sense but not in a material form, may be excluded, or one may restrict the choice of the design of a certain component of the object to a certain class of designs which are already known from the past. So it is important to notice that W does not contain all models which may be thought of in a theoretical sense. The formal descriptions in W are models according to the signature of the language of the design requirements, or extensions of that language. The class W serves as a domain from which possible solutions of a design process may be chosen. Since taking expansions is also a variable in the design process, W is multiform with respect to signature.

In many cases there exists some minimal language L which is interpreted in all these models (containing the language of the initial design requirements). Then the reducts of all models in W with respect to L may be taken; they form a class W_0. In the example of Table 1, one may think of W_0 as the class of models of the form $<R, c>$ for all $c \in R$ and for W one may take W_0 extended by all models of the form $<R, c, d>$ for all $c, d \in R$. For the general case we do not assume that W_0 is a subclass of W; this means W may be not closed under taking reducts. By P we denote the class of all partial models involved in the design process; we assume $W \subset P$. In these partial models some symbols of the language of design requirements may have no interpretation, or only a partial interpretation. In the example of Table 1, for P we may take W extended by the partial models $<R, c, *>$, $<R, *, d>$ and $<R, *, *>$ for all $c, d \in R$.

One could think of the condition that every $M \in P$ has some refinement M^* in W, i.e. every partial model may be filled out and realised. However, to leave open the possibility of a negative outcome of the design process (asserting that there does not exist an object satisfying the requirements) we de not assume this condition. This means we may reason with partial models of inconsistent theories.

2.6. General domain theory of the objects

The objects from W share certain general properties or laws, for instance physical laws. The knowledge about these is called the *(general) domain theory of the objects.* This theory is denoted by T. Every $M \in W$ is a (possibly partial) model of T, i.e. every expression from T which may be interpreted in M is true in M. We assume, moreover, that also every partial model from P is a model of T.

During the design process, T may be used to derive information on the partial model M at hand. For instance, if T contains a Horn proposition of the form $a \wedge b \rightarrow c$ and a and b are both true in M and c is not filled in, then M may be refined to a partial model in which also c is true. This again gives an assumption on P, viz. such a refinement may indeed be found inside P, unless there is an inconsistency. The case that inconsistencies occur will have to be considered more closely; at this moment we leave this open. In the example of Table 1, T consists of all the well known axioms for calculations with real numbers, such as associativity, etc. This general knowledge is used frequently during the design process as sketched in Table 1. Note that W may not contain all models of T; for instance W may be restricted to models built up from known components. One could try to express such restrictions in the form of knowledge in T, but this may be hard.

Summarising, we have the following logical description:

L: a language of a certain signature

T: a theory of objects in a language extending the language L

W: a class of models of T of different signatures extending the signature of L; these are descriptions of realisable objects

P: a class of partial models of T of different signatures extending the signature of L; these are incomplete descriptions of objects being designed; W is a subclass of P.

W_0: the class of reducts of the models of W to the language L

2.7. Logical descriptions of refinement steps

The process of refinement often plays a central role in design. Therefore, we summarise the logical techniques which may be used to describe refinement steps:

- extending the language
  – propositional logic: adding new atomic propositions
  – predicate logic: adding new constants, functions, predicates
- extending a model
  – propositional logic: assigning truth values to more atoms
  – predicate logic:
  i. extending existing, partially defined functions or predicates
  ii. assigning (partially) interpretations to new constants, functions or predicates

The question may be raised whether or not the predicate-logical (model-theoretic) notion of model is adequate for representing the type of structures as used in design. In this paper this question is left open. A related question is whether for different types of design tasks (such as architectural design, mechanical design, program design) different notions of models should be used. If the answer to the latter question is affirmative, then as an underlying (meta-)question could be investigated: what kind of properties the notion of model should possess.

A number of special cases of refinement processes may be distinguished (for instance, see [8]); we mention:

incremental refinement across abstraction levels. Here the domain is structured according to a number of abstraction levels; the design process proceeds by first determining solutions on the more abstract levels and then expanding these solutions incrementally to the more concrete levels. An example of this form may be found in §5 below.

hierarchical decomposition. In this case the object is built up from a number of components which may be designed separately (also called top-down design) and put into some interaction afterwards. Of this strategy too, an example may be found in §5.

3. The Theory of Requirements; Reasoning from and About this Theory

The logical description of the design objects in various measures of detail as given above leaves open the role of the design requirements. These design requirements give an implicit (and often incomplete) description of the object being designed; this may be compared with an equation which gives an implicit description of a certain type of number. The aim of the design process is to use this implicit information to construct an explicit description of one or more of the objects satisfying the design requirements (construct one or more solutions of the equation).

On the one hand the design requirements play the role of a logical theory for which a model is asked (for instance, see the example in Table 1). From this logical theory conclusions may be derived. On the other hand we also noticed that the set of requirements is changing dynamically during the design process. At some stages of the design process some other (sub) design problems are solved with different sets of requirements. Major design strategies are based upon some idea of splitting up the design problem into subproblems; each of these subproblems has its own set of design requirements. The sets of requirements of these subproblems are derived in some way or other from the initial design requirements (transforming them, choosing a suitable subset, etc.). The reasoning for such a derivation can only be described by considering the theories of requirements on a meta-level: one has to reason *about* theories. Sometimes such a form of reasoning is called *reasoning across viewpoints* (see [1]). We go into these two roles of the set of design requirements in some more detail.

3.1. The set of design requirements as a logical theory

To start with we consider a set of design requirements as a (finite) *logical theory* R which need to be satisfied by the object being designed. In this sense the theory of design requirements supplements the domain theory T of objects as defined above. The design process succeeds if some explicit model of the theory $R \cup T$ has been constructed. The design process has a negative outcome if it is proved that such a model does not exist, i.e. the theory $R \cup T$ is inconsistent. The partial models used in the various stages of the design process are partial models of the theory of design requirements.

If during the design process a partial model M is considered, the theory R may be used to derive information about M; this is similar to the use of T as described above. For instance, if R contains the disjunction $a \vee b$ and in M we know that a is false and b is undecided, then M may be refined to a partial model in which b is true. In the example of Table 1, such reasoning from the design requirements is used to derive from $x + 2 = y$ and $y = 3$ the statement $x = 1$. Therefore, a refinement of the partial model may be created where x is given the value 1.

Below we first try to sketch a clear picture of the kind of reasoning that is intended in our framework (in §3.2). In §3.3 we study the role of the theory of design requirements in reasoning in some more detail.

3.2. The dynamics of reasoning

In the above we gave examples of reasoning from the domain theory of objects (in §2.6) and reasoning from the theory of design requirements (in §3.1). It may be noted that this provides a *dynamical description* of reasoning as *making inferences* followed by *refining the partial model* at hand. This means that the result of the reasoning is not a true statement but a more detailed description of the object involved. According to this view the reasoning may be described dynamically by a trace of inference steps combined with a trace of partial models. If we permit the possibility of partial models which cannot be refined to a complete model in W, sometimes the refinement step in the reasoning may be impossible. In that case backtracking to an earlier choice point is required. In this sense the dynamical view of reasoning we suggest supports nonmonotonic reasoning. It may even be the case that the theory $R \cup T$ is inconsistent; in that case it only has partial models.

Reasoning from a fixed theory is not the only kind of reasoning that is done. As noted earlier the theory of design requirements changes dynamically during the design process. In fact, to use the implicit information embodied by the requirements two different types of reasoning are employed:

- reasoning *from* the domain theory and the theory of design requirements; (combining knowledge from the theory with information from the partial model to new information which is used to refine the partial model)
- reasoning *about* the theory of requirements; (meta-reasoning where the requirements are used as objects to reason about, represented by terms in the meta-language and for which variables may be used)

The first of these two types of reasoning is described in §2.6 and §3.1 above. The latter type of reasoning will be illustrated below in §3.3 by a number of specific forms. In §3.4 we will discuss the interaction between the two types of reasoning.

3.3. Reasoning about the theory of design requirements

Four specific uses of reasoning about design requirements may be mentioned; this is not an exhaustive list:

- transforming the theory of requirements into an equivalent theory in the same language
- translating the theory of requirements into a theory in a different language
- first choosing a part of the theory of requirements
- hierarchical decomposition
- adding a specific requirement to the theory of requirements (making a choice on which requirement to add)

We will discuss these below.

3.3.1. Transforming into an equivalent theory. The first specific type of reasoning we consider is the transformation of the theory of design requirements into a different but equivalent theory, i.e. a theory with the same class of models. Transforming a set of equations into an equivalent set, as occurs in some steps in Table 1 is an example of this. In these cases it is important to be able to describe the connection between the two theories and to reason with that connection. Therefore a meta-language is needed in which one can speak about theories, models, satisfaction and derivability (for instance see [2]). In the case of a set of requirements R_1 which is transformed into an equivalent set R_2 the connection between them may be expressed as:

$$\text{true}(M, T) \Rightarrow [\text{true}(M, R_1) \Leftrightarrow \text{true}(M, R_2)]$$

Here the atomic meta-statement true(M, A) means that the model M is a model of the theory A. This is a semantical assertion which may be derived (using soundness) from provability relations like

$$\text{provable}(T \cup R_1, R_2) \wedge \text{provable}(T \cup R_2, R_1)$$

One may reason with the meta-knowledge given above. For instance such meta-reasoning is needed to derive that a solution found for R_2 is also a solution for R_1. A knowledge based design system should be able to make such inferences.

3.3.2. Translating into a theory in a different language. It may happen that the theory of requirements is transformed into a theory in another language. In this case the connection between the two theories does not quite conform to the description in §3.3.1. For instance, in the example in Table 1 the step from I to II consists of translating the theory $R_1 = \{x^2 + 4x - 5 = 0\}$ into the new theory $R_2 = \{y^2 - 9 = 0\}$ which is formulated in a different language. The model with x is translated into a model with y. Here the connection between the two theories may be expressed by taking the translation relation $I = \{x + 2 = y\}$ as an additional condition:

$$\text{true}(M, T \cup I) \Rightarrow [\text{true}(M, R_1) \Leftrightarrow \text{true}(M, R_2)]$$

A different way to express the connection is by considering the two extended theories $I \cup R_1$ and $I \cup R_2$. The connection between these extended theories may be expressed by

$$\text{true}(M, T) \Rightarrow [\text{true}(M, I \cup R_1) \Leftrightarrow \text{true}(M, I \cup R_2)]$$

In both cases we deal with expansions with respect to the initial theory R_1.

3.3.3. First choosing a subset of the set of requirements. In the example of Table 1 the condition $x \geq 0$ has been left out for some time. This strategy is often used in design processes: first find a solution for a part of the set of requirements, then refine this solution to a solution for the complete set of requirements. In this case the theory of requirements R is split up into two new theories: the chosen subset R_1 and the set of remaining requirements R_2. Here the connection may be expressed as:

$$\text{true}(M, T) \Rightarrow [\text{true}(M, R) \Leftrightarrow [\text{true}(M, R_1) \wedge \text{true}(M, R_2)]]$$

In all of the three specific types of reasoning about theories of design requirements considered until now, there turns out to be some form of equivalence. This is not the case for hierarchical decomposition, as described below.

3.3.4. Hierarchical decomposition. Often an object is designed by building it up from a number of components; these components may be built up from subcomponents, and so on. This is a well known strategy in designing called *hierarchical decomposition* (for instance see [8]). For each of the components a new theory of design requirements will have to be created. This new theory of requirements defines a new design problem, a subproblem of the initial design problem. The design process may proceed by solving this subproblem first (top-down strategy). Furthermore, an additional theory of design requirements will have to be created to describe the interactions between the components. Since the whole construction is intended to provide a solution for the initial theory of requirements there must exist some logical connections between all these theories. These logical connections may be described as follows.

Suppose R is the initial theory of design requirements and we decompose into two components with theories of requirements R_1 en R_2. Let I be the theory of requirements which describes the interaction between the components and let M be an arbitrary model, built up from the components M_1 and M_2. Then the logical connection between the theories may be expressed by

$$\text{true}(M, T) \,\&\, \text{true}(M_1, R_1) \,\&\, \text{true}(M_2, R_2) \,\&\, \text{true}((M, M_1, M_2), I) \Rightarrow \text{true}(M, R)$$

Here the fourth conjunct expresses that in M the components M_1 and M_2 have an interaction which satisfies I. This will have to be described more precisely. The semantical description above is implied by a provability relation of the form:

$$\text{provable}(R_1' \cup R_2' \cup I \cup T, R)$$

Here the primes indicate that the requirements of R_1 and R_2 should be taken as relative with respect to the component concerned.

The logical description of hierarchical decomposition, as well as the other specific transformations in §3.3.1 to §3.3.3, shows us that reasoning about theories (or viewpoints) plays a major role in design processes.

3.3.5. Adding a specific requirement; generate-and test. Sometimes the design process will proceed by making a choice for some specific aspect of the object being designed. In this case a subclass of the class of possible models is tried first. If this subclass does not contain a solution, backtracking is necessary to where the choice was made. The well known method called *generate-and-test* (for instance see [8]) makes use of such choices. A specific form is obtained if the additional requirement gives explicit information about the model.

From the logical point of view adding a specific requirement means creating a stronger theory which may be inconsistent (the case that backtracking will occur). As a logical peculiarity such an inconsistent theory is used not only to reason about but also to reason from. The logical connection between the two theories is a simple one:

$$\text{true}(M,T) \Rightarrow [\,\text{true}(M,R_2) \Rightarrow \text{true}(M,R_1)\,]$$

As in the case of hierarchical decomposition there is no equivalence; therefore backtracking may be necessary.

3.4. The interaction between object-reasoning and meta-reasoning

In the §3.1 to §3.3 above we described independently the two types of reasoning involved in our logical framework: object-reasoning (reasoning from the domain theory and the theory of design requirements) and meta-reasoning (reasoning about theories of design requirements and partial models). In a design process these two types of reasoning are mixed to a complex pattern of reasoning. To give a complete logical description of such a complex pattern of reasoning it is important to give logical descriptions not only of the types of reasoning themselves but also of the interactions between them.

For a logical description of the interaction between the object-reasoning and the meta-reasoning we make use of *reflection principles* similar to [14]. These are transformations of object knowledge to meta-knowledge *(upward)* or vice versa *(downward)*. The conclusion A of a derivation from the domain theory and the theory of design requirements $\cup$ R using data from the partial model M is transformed by an upward reflection principle to the meta-datum that A is provable from T $\cup$ R and data from M. This reflection principle may be formulated as follows:

$$\frac{T \cup R \models_M A}{\text{provable}(T \cup R,\ M,\ A)}$$

A similar upward reflection principle may be formulated for partial models:

$$\frac{M \vdash T \cup R}{\text{true}(M,\ T \cup R)}$$

The upward reflection principles may be used to provide the meta-reasoning with meta-data. The downward reflection principles are the converse of the upward ones.

An example of the use of the reflection principles in the whole design process may be sketched as follows. First the theory of design requirements R is transformed to another one R′ by meta-reasoning (for instance according to §3.3.1). Then a partial model is chosen for this transformed theory of design requirements. This partial model is refined to a complete model M by object reasoning. This fact is reflected to the meta-datum

$$\text{true}(M,\ T \cup R')$$

By meta-reasoning it is derived also that

$$\text{true}(M,\ T \cup R)$$

Finally, by the downward reflection principle it is concluded that M satisfies T $\cup$ R. Here object-reasoning may continue. Note that we did not go into the strategy of reasoning; we will say a word about that in the section below.

4. The Design Strategy

In §2 and §3 we sketched a number of logical connections between the theories and models before and after a step in the design process. These logical connections are described as logical inference steps on the object-level or on a meta-level or as upward or downward reflections. A complete survey of the steps which are possible and correct from the logical point of view is a *basic layer* of description; this layer still has to be described more exhaustively.

We may already say, however, that the search trees produced at this level of description will tend to be huge. To avoid exhaustive search an additional *strategic layer* is needed which contains knowledge about the inference strategy of the basic layer and can reason about the strategy. For instance in Table 1 for each of the steps it is not argued why it is chosen instead of the many alternatives that are possible. The strategic layer is implicit in this example.

The content of the strategic layer depends heavily on the kind of design process. In routine-design, where a fixed structure of the object is given, only the components have to be specified explicitly. In a more creative kind of design the structure itself is also left open. In [3] these kinds of design processes are described in some more detail, and the case of routine-design is treated more extensively. In the case of routine-design the strategy may be expressed by an explicit algorithmic description for the steps to be taken. This may be compared to the solving of an equation by someone who has learned a fixed algorithmic schema for it. The more creative case may be compared to someone who tries to solve an equation for the first time, without having been told how to do it. In this case a different more general problem solving strategy is needed.

In this paper we only mention the fact that such a strategic layer has to be worked out; we do not give suggestions for this. In [8] the two layers (basic and strategic) considered here are called 'knowledge level' and 'functional level'.

5. An Example: Making a Floor Plan for a House

This example is described in an informal manner in [8], p.136/137; we give a formalised description of a simplified form. Our intention is to show how the framework described in the earlier sections applies in more realistic situations. The example deals with a single-story house of rectangular form. For example as in Fig. 1.

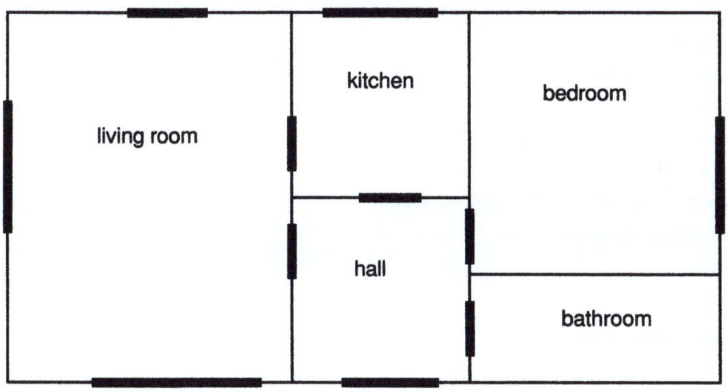

Fig. 1. Floor plan of a house

5.1. The logical language

In the logical language it should be possible to express connections between rooms such as "room x shares a door with room y", or "room x shares a door or a window with outer wall z". Further one should be able to express that certain dimensions satisfy certain restrictions: "wall x of room y has a length less than some value".

To realise these expressivity requirements we define the logical language as follows. The objects to talk about are partitioned into the sorts rooms, walls, corners, natural numbers. There are constants to name specific rooms: k (kitchen), b (bathroom), l (living room), s (bedroom), h (hall), w (the whole house) Every room has four corners, defined by the functions cor1, ⋯, cor4 from the sort of rooms to the sort of corners. The corners are numbered from left-front anti-clockwise to left-back; they have coordinates defined by the functions x–co, y–co from the sort of corners to the sort of natural numbers (taken in decimetres). Every room has four walls: front, back, left, right; the same holds for the whole house. These are defined by four functions from rooms to walls, namely lw, rw, fw, bw. Besides a predicate wall on walls × rooms may be used to denote that a certain wall is one of the walls of the given room. The symbol length denotes a function from walls to natural numbers (the length of the wall in decimetres). The natural numbers have the usual ordering, addition and subtraction.

The predicate door on walls × walls is used to denote sharing a door, and the predicate window on walls × walls to denote sharing a window. Some function overlap is needed from walls × walls to natural numbers to denote the overlap of two walls. For a complete description some more functions and predicates may be needed. For instance predicates to express that two rooms are disjoint. For reasons of exposition we do not give all technical details for that; we confine ourselves to a subset of the specification. Summarising, we define the following signature:

- sorts:
  rooms, walls, corners, natural numbers
- constants:
  k , b , l , s , h , w in rooms
- functions:
  lw, rw, fw, bw : rooms → walls
  co1, co2, co3, co4 : rooms → corners
  x–co : corners → natural numbers
  y–co : corners → natural numbers
  length : walls → natural numbers
  +, − : natural numbers × natural numbers → natural numbers
  overlap : walls × walls → natural numbers
- predicates:
  wall on walls × rooms
  ≤ on natural numbers × natural numbers
  door on walls × walls
  window on walls × walls

Next we describe the general domain knowledge, the theory of the objects.

5.2. Theory of objects

The theory T consists of the following conditions; these are the properties shared by all design objects. This list is not intended to be complete.

wall(fw(x), x) wall(bw(x), x) wall(lw(x), x) wall(rw(x), x)	Here it is summed up that front wall etc. of a room are specific cases of a wall of that room.
door(x, y) → overlap(x, y) ≥ 12 window(x,y) → overlap(x, y) ≥ 17	Doors and windows should have certain minimal dimensions.
length(x) ≥ 15	The walls should have certain minimal dimensions

x–co(cor1(x)) = x–co(cor4(x)) x–co(cor2(x)) = x–co(cor3(x)) y–co(cor1(x)) = y–co(cor2(x)) y–co(cor3(x)) = y–co(cor4(x))	The rooms are rectangles.

Additional knowledge should be added that the rooms do not overlap and together form the whole house. As mentioned earlier we do not go into the technical details.

5.3. Theory of design requirements

The design problem we consider may be described by the following design requirements; they form the theory R:

$\mathrm{length(lw(w))} \leq 70$ $\mathrm{length(fw(w))} \leq 100$	These are the maximal permitted dimensions for the house.
door(lw(k), lw(w))	The kitchen should have an outside door in its left wall.
door(bw(w), bw(w))	The living room should have a back door to the garden.
door(fw(h), fw(w))	The hall should have a front door.
window(fw(w), fw(w))	The living room should have a window in the front wall of the house.
$\exists$ x,y wall(x, s) $\wedge$ window(x, y)	The bedroom should have a window.
$\exists$ x,y wall(x, k) $\wedge$ wall(y, l) $\wedge$ door(x, y)	The kitchen and the living should be connected by a door.
$\exists$ x,y wall(x, h) $\wedge$ wall(y, l) $\wedge$ door(x, y) $\exists$ x,y wall(x, h) $\wedge$ wall(y, s) $\wedge$ door(x, y) $\exists$ x,y wall(x, h) $\wedge$ wall(y, b) $\wedge$ door(x, y)	The hall should be connected by a door with living, bedroom and bathroom.

If necessary, the existential quantifiers may be replaced by additional (Skolem) constants.

5.4. Two examples of design strategies and their logical description

In the floor plan example the following two design strategies may be useful (see also §2): *incremental refinement using different abstraction levels* and *hierarchical decomposition.* We sketch a logical description for both.

5.4.1. Incremental refinement using different abstraction levels. This may be thought of as first solving the design problem in a qualitative, topological manner (connections between the rooms) and subsequently refining this qualitative solution to a quantitative, metric solution (the exact sizes).

From the logical point of view this means that first a restricted language is used, containing only qualitative expressions, i.e. all expressions containing numbers are deleted from the theory R. For the language given above this is realised by deleting the sort natural numbers and the operations on it and the functions x–co, y–co, length and overlap. For this reduced theory a model is constructed; this is a partial model for the theory R. Subsequently this partial model is refined to a complete model of the theory R.

5.4.2. Hierarchical decomposition. This strategy may be applied by splitting up the house in two main components (kitchen/living room and hall/bedroom/bathroom) and further splitting up these components into the different rooms, as in Fig. 2. A logical description of this strategy requires an expansion of the theories T and R to be able to express the hierarchy of Fig. 2 (for instance predicates for the two main components should be added and a predicate for the decomposing relation). A further logical description may be obtained by working out the general description given in §3.3.4. By the axioms for the theory I for instance it has to be expressed that the size of the one component together with the size of the other component is just the size of the whole house (which is subject to certain restrictions).

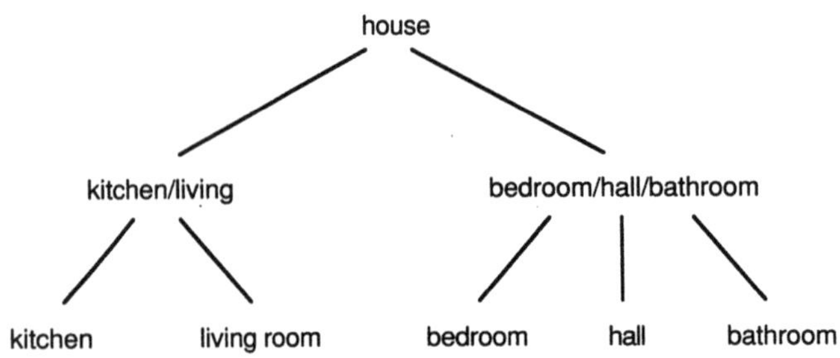

Fig. 2. Hierarchical decomposition for designing a floor plan

6. Conclusions

In the above sections we sketched logical descriptions of the steps involved in a design process. The logical means used for these descriptions are the following:

- partial models to represent the object being designed on different levels of detail
- the domain theory of design objects; partial models are models of this theory
- the theories of design requirements; these theories vary during the design process
- reasoning from these theories, while making use of data from the partial model at hand
- reasoning about viewpoints (theories of design requirements), while making use of (meta) data from the viewpoints at hand

The use of partial models as proposed here is consistent with the use of partial models to represent changing situations in [11]. Reasoning from a given theory while making use of data from the model at hand is described in more detail in [8]. Reasoning about viewpoints and the interaction with other kinds of inference processes is similar to the descriptions studied in [9] concerning diagnostic processes and in [11] concerning process control.

It may be clear that representing the knowledge and the reasoning involved in a more realistic design process is not a simple issue. A number of aspects have not been considered above: for instance the strategy of making a choice for the dimensions of a certain room and then trying to adapt the other rooms. Some of these aspects are described in [8].

Acknowledgements

Earlier drafts of this paper have been read and commented upon by Pieter Geelen, Piet Rodenburg and Roel de Vrijer (Programming Research Group, University of Amsterdam). Also some discussions have taken place with Paul ten Hagen and Paul Veerkamp (IIICAD project, CWI, Amsterdam), with Leo van Geffen and Lucienne Stomp (Engineering Design Laboratory, University of Twente), and with Tetsuo Tomiyama, Micheal Mac an Airchinnigh and others at the ICAD-workshop at Texel, April 1989.

References

1. G. Attardi and M. Simi, "Metalanguage and Reasoning Across Viewpoints," *Proceedings ECAI*, 1984.
2. K. Bowen and R. Kowalski, "Amalgamating Language and Meta-language," in *Logic Programming*, ed. Clark and Tarnlund, Academic Press, London, 1982.
3. D.C. Brown and B. Chandrasekaran, *Design Problem Solving; Knowledge Structures and Control Strategies,* Pitman, 1989.
4. P.J.W. ten Hagen and T. Tomiyama (eds), *Intelligent CAD Systems I: Theoretical and Methodological Aspects,* Springer-Verlag, 1987.
5. J. Mostow, "Toward Better Models of the Design Process," *AI Magazine*, pp. 44-57, 1985.
6. C.S. Ocathain, "Why is Design Logically Impossible?," *Design Studies*, no. 3, pp. 123-125, 1982.
7. T. Tomiyama and H. Yoshikawa, "Extended General Design Theory," *Proceedings of the IFIP WG 5.2 Working Conference on Design Theory for CAD*, pp. 95-125, North-Holland, Amsterdam, 1987.
8. C. Tong, "Toward an Engineering Science of Knowledge-based Design," *Artificial Intelligence in Engineering*, no. 2, pp. 133-166, 1987.
9. J. Treur, "On the Use of Reflection Principles in Modelling Complex Reasoning," Report P8812, Programming Research Group, University of Amsterdam, 1988.
10. J. Treur, "Design of Modular and Interactive Knowledge-based Systems," Report P8826, Programming Research Group, University of Amsterdam, 1988.
11. J. Treur, "Reasoning about Partial Models, Actions and Plans," Report P8813, Programming Research Group, University of Amsterdam, 1988.
12. J. Treur, "Completeness and Definability in Diagnostic Expert Systems," *Proc. European Conf. on Artificial Intelligence, ECAI-88*, Munchen, 1988.
13. R. Turner, *Logics for Artificial Intelligence,* Ellis Horwood, Chichester, 1984.
14. R.W. Weyhrauch, "Prolegomena to a Theory of Mechanized Formal Reasoning," *Artificial Intelligence*, no. 13, pp. 133-170, 1980.

2

A Psychologically Based Intelligent Design Aid

Keith Baker[†], Linden Ball, Phil Culverhouse, Ian Dennis, Jonathan Evans, Peter Jagodzinski, Pat Pearce, Dean Scothern, and Gill Venner

Department of Computing
Plymouth Polytechnic
Plymouth PL4 8AA, UK

Abstract: This paper reports on an approach to intelligent design aiding in electronic engineering that differs from both (a) conventional CAD approaches aimed at developing draughting tools to assist with low-level circuit design and simulation, and (b) expert system approaches aimed at automating features of the electronic design process. By contrast, the system that we are developing is designed to work cooperatively with engineers, facilitating their thought processes and helping optimise the designs produced, particularly in the early, high-level stages of design. To develop such a system clearly requires an understanding of the cognitive processes involved in design which can only be achieved by systematic psychological studies. Accordingly, the research described here has involved a collaboration between experimental psychologists, computer scientists and electronic engineers who have been jointly engaged in the development of the prototype design aid which we call the Plymouth Engineer's Design Assistant (PEDA). The primary aim of the psychological work has been to develop a model of the design process. The model that we have derived to date has strongly influenced the construction of the PEDA both in its functionality and in the specification of the user interface. A designer is able to interact with the system via a direct manipulation interface which offers a drawing board metaphor. Any part of the evolving hierarchical design representation can be accessed and the functionality that is supported ranges from mathematical functions to specific electronic components. Multiple versions of all or part of the design are maintained by the PEDA, and a history of design steps and decisions establishes a model of the designer's use of the tool. Within this paper we also discuss some future directions of our work.
Keywords: Intelligent CAD, Electronic Design, Psychological Model, Cognitive Processes, Hierarchical Design Representation, Multiple Versions, Direct Manipulation Interface.

1. Introduction

The human creative process of electronic circuit design is a complex iterative activity drawing upon multiple sources of knowledge derived from prior experiences. The designer seeks to optimise the required functionality of an artifact within the constraints of the bounding external

[†] Present address: Dept. of Computer Science, University of Reading, Reading, Berkshire, RG6 2AX, UK

parameters. The early or conceptual stage of the process is dominated by the generation of ideas which are subsequently evaluated against general requirements' criteria. There follows a process whereby additional data are incorporated allowing decisions to be made between competing alternatives as more tangible evidence of function is derived. Once a final selection is made the design process becomes a routine transformation of function into the low level, circuit implementation detail.

Many, if not all, commercially available electronic CAD systems are merely sophisticated draughting tools that assist with this final stage of circuit level design. Few systems address the early stages of the design process where, to be a useful adjunct to human performance, the tools must closely match the cognitive processes associated with the various design activities. A psychological model of the human designer is therefore an essential prerequisite to any successful implementation of a computer based design assistant. Accordingly, the research reported involves a collaboration between a team of electronic engineers, computer scientists and cognitive psychologists with the aim of developing an intelligent software aid for the facilitation of engineering design. The work completed to date includes a psychological study of the cognitive processes involved in engineering design and the development of a prototype software system, within the Sun/ART environment, known as the Plymouth Engineer's Design Assistant (PEDA).

The PEDA system is intended to provide support for the designer in the higher levels and earlier stages of engineering design, in contrast to conventional CAD systems which provide low level tools for use at a stage when the major planning and structuring decisions have already been taken. Hence the system can be viewed at one level as an interactive decision aid which develops a model of the user and his/her problem and facilitates the formulation and assessment of alternative scenarios or versions of the design solution. The system is also intended to provide more specific and lower level aid in the form of direct simulations of design concepts and the inclusion (eventually) of relevant databases with intelligent search routines and rule bases.

The system we are developing is designed to take account of an understanding of the psychological characteristics of the engineering design process and any attendant cognitive limitations and weaknesses for which the system may compensate. Hence, our work is distinguished from most other approaches to engineering design aids in that it is firmly based upon psychological study of the mental processes involved. The cognitive psychologists in the group have applied their theoretical knowledge of the processes of thinking and decision making to a systematic study of the thought processes used by engineers. In view of the fundamental importance of the psychological analyses of design processes to the PEDA system, we start with a discussion of this aspect of the research.

2. Psychological Analysis of the Engineering Design Process

Psychologically oriented research on engineering design remains extremely scarce, and that which has been undertaken tends to have been very narrowly focused on possibly unrepresentative design tasks (e.g. [6, 22]). Psychological research dealing with other design domains such as architectural design and software design, whilst somewhat more plentiful, may clearly have limited applicability to the engineering domain (see, for example, studies by Eastman [7, 8]; Jeffries et al. [13]; and Adelson & Soloway [1]). In contrast to the paucity of knowledge derived from rigorous psychological analyses of design, an abundant speculative literature is available – written by practitioners – that claims to shed light on design processes and which usually advocates methodologies and formal procedures which should be adhered to for the production of enhanced designs (see, for example [2, 14, 19]). It is felt, however, that such intuitive and conjectural information provides an inadequate basis for the development of interactive design environments, particularly in the light of abundant psychological evidence which suggests that people – experts included – have very poor self-knowledge of their higher cognitive processes (e.g. those involved in problem solving and decision making) despite belief that they know the strategies and procedures that they use (see [11] for a detailed discussion). The point is, then, that whilst self-knowledge is not required to perform well at a task, such knowledge is required if you want to describe the methods and procedures that you use for the benefit of other people. Indeed, the lack of access to such procedural knowledge presents one of the major problems for orthodox expert systems approaches to intelligent design aiding based on conventional knowledge engineering techniques such as extended interviewing.

The formulation of our understanding of engineering design processes has therefore been based upon:

1. Review and assessment of the limited directly relevant literature referred to above,
2. Consideration of general theoretical principles established in the wider psychological literature on thinking and problem solving, and
3. The conduct of our own empirical studies of cognitive processes in engineering design.

The psychological objectives of these studies have been:

1. To discover the nature of any general design schemas or strategies utilised by engineers, and
2. To determine the nature of any cognitive limitations (e.g. working memory capacity, judgemental biases) which constrain the attainment of effective designs.

Such understanding of the underlying cognitive processes has direct implications for the development of an intelligent design aid that allows the engineer to design flexibly and naturally while at the same time helping to counteract cognitive failures (e.g. by problem structuring, reminding, advising and guiding). It should be emphasised that CAD systems derived from intuitive models of design processes could conceivably hinder rather than assist a designer's creative work. Such hindrance could be caused, for example, by the system enforcing adherence to an unnecessarily rigid sequence of design activities or imposing added demands on cognitive resources at points when the designer's cognitive system is already strained.

One exploratory study of the design process that has been undertaken focused on eight semi-expert electronic engineers (final year undergraduates) tackling individual long-term design and development projects. All the engineers were pursuing projects within their favoured subdomain of electronic engineering and had previously undergone a period of placement in professional design settings. Regular structured diary entries and intermittent interviews were obtained from subjects over a period of several months during which the projects were run. This information therefore related to activity occurring throughout a design project – from the foremost stages of conceptual design to the final stages of realising a prototype in hardware. The protocols acquired in this study were subjected to detailed qualitative analysis using a technique similar to that devised by Byrne [4, 5] for studying verbalisations produced by subjects tackling *mental planning* tasks (e.g. designing the the interior layout of an apartment). As with Byrne's technique, our method was primarily concerned with identifying the organisation and sequencing of subjects' activities (as conveyed through their verbal reports) rather than with a simple examination of the surface content verbalisations. Particular consideration was given to the goals and subgoals which were motivating design activity, though our analysis also looked at other features of the designers' behaviour, for example, the nature of evaluation processes such as those involved in assessing the viability of a solution concept or those used for selecting one design option from a set of alternatives. The question of what type of evaluation procedures are used by engineers in these latter kinds of situations is of clear psychological importance and has obvious relevance to the development of a design aid that can lend appropriate assistance.

The results of our protocol analysis indicated that subjects' initial processing effort was directed toward

1. Determining the functional requirements of the intended artifact (i.e. what the artifact was required to do)
2. Detailing high-level constraints relating to the artifact's performance, resource usage and style (e.g. efficiency, reliability, testability, power consumption, speed and the like), and
3. Defining constraints on the design process itself (e.g. budget, time allocation and equipment availability).

Subsequent to this initial phase of processing, subjects were seen to use a *problem reduction* strategy in their continuing design work. In the first instance this strategy led to a division of the overriding design problem into a collection of manageable and minimally interacting subproblems concerning the design of separate functional modules of the desired artifact. Once identified, the subproblems were then focused on in an essentially sequential manner with functional modules being developed depth-first to completion. In many cases subjects tended not to expend cognitive effort in the search for and modelling of substantially alternative (and potentially optimal) design solutions. Instead they operated a *satisficing* principle, focusing selectively on a single satisfactory high-level solution concept rather than comparing alternatives with the aim of optimising choices. That is to say, subjects seemed to accept solutions that were *good enough* (e.g. cheap enough to implement or fast enough in operation) rather than looking for solutions that, even if not *best* (i.e. cheapest, fastest), were at least somewhat *better* than the first solution generated. It is possible, of course, that these subjects were engaging in some form of rapid and covert exploration and evaluation of alternative solutions, although such intuitive processing clearly has an air of superficiality about it. Indeed, if this were the case then it is possible that encouraging a more explicit mode of thought in which options are more thoroughly modelled and assessed might lead to enhanced design performance.

The notion of satisficing was introduced by Simon [20] who advocated it as the basis of an acceptable procedure for finding a satisfactory design solution in the absence of a method for finding an optimum. Now while this argument is reasonable enough, it seems certain that in a competitive, profit-oriented design climate, designs that are merely satisfactory are rarely going to prove to be a truly cost-effective solution. Clearly, then, a design system that facilitates the search for and evaluation of alternative design concepts might be of much value to contemporary designers who generally are required to produce as optimal a design as possible within as short a time as possible.

In the present study it was additionally observed that whenever a designer selected a satisfactory high-level solution concept, this concept would often then be developed gradually to completion by the production of a variety of slightly improved versions. This suggests that the rudiments of an optimisation strategy was possessed by these engineers and emerged at lower levels in the design hierarchy. Considerable evidence was also derived in this study which indicated that subjects were constructing and manipulating mental models to simulate the dynamic behaviour of aspects of the developing design [1, 15]. Such modelling would seem to provide a potentially important means for comparison and evaluation of the viability of alternative solutions to a design problem or subproblem.

The main outcome of the study described has been the derivation of a theoretical model of the global processes that control and coordinate the designer's mental movement between subproblems and his/her development and evaluation of design solutions. This model incorporates the idea that a designer possesses generalisable abstract knowledge concerning how to produce good designs in the form of a *design schema* [13]. This high-level schema is built up through experience and can be applied to a wide range of domain problems having similar fundamental structures but which differ in terms of their content. When applied in a design situation the design schema controls the decomposition of the design problem into subproblems relating to the design of component functional modules. The schema then coordinates the order in which these subproblems are worked on and the search, retrieval and evaluation of solutions as well as initiating further subproblem decomposition if this is necessary (in which case the design schema is invoked recursively). The basis of the design schema is clearly the problem reduction strategy outlined above. Problem reduction is a general purpose, domain independent, problem solving method – one of the so-called weak methods of Newell and Simon [18]. When, however, the problem reduction method is augmented with procedures that ensure the effective retrieval and utilisation of domain specific technical knowledge, the outcome is a sophisticated higher-order knowledge structure that can function to produce fairly expert solutions to a wide range of design problems. In our theoretical model of design we have summarised the major processes of the design schema as a set of abstract production rules. A production rule representation of high-level design knowledge seems to provide a useful means to capture the flexibility of design activities as well as their fundamental structuredness.

To conclude the discussion of this study we would like to stress that it was very much an exploratory attempt to penetrate into a fairly unchartered area of psychological interest. In light of the inductive nature of the research we are aware of a definite need to be cautious in generalising the results to design situations involving (i) expertise levels different to those of the subjects studied and (ii) problems having characteristics different to those undertaken in the present case. A further problematic aspect of the study relates to the data collection methods that were adopted – in particular the use of cognitive diaries. Whilst this was the most practical solution we could think of for studying thought processes extended over many months, it clearly has its limitations. For example, subjects may sometimes have presented rationalised accounts of their behaviours or at other times may simply have forgotten to mention important design activities. Having voiced the need to treat the results of this study in a tentative manner we must emphasise that we remain particularly encouraged by the distinct commonalities that were observed in the design styles of the individuals studied who were engineers of varying ability pursuing designs in a range of technological subdomains.

In an attempt to further our understanding of engineering design processes we have recently performed another investigation, this time using a more sophisticated method for studying ongoing activities as well as focusing on engineers with a considerable level of design experience and expertise. This second study involved the collection of concurrent video-protocols of individual professional engineers *thinking aloud* as they attempted a small-scale design task in a laboratory setting. Whilst less realistic than the longitudinal study described above, this think-aloud (TA) method affords the considerable advantage of providing a continuous and concurrent record of a specific act of thought permitting more accurate study of the underlying cognitive processes.

The most powerful theoretical arguments concerning the value of TA protocols in cognitive research have been provided by Ericsson and Simon [9, 10]. These authors argue that concurrent verbal reports provide details of the information attended to or heeded by the subject at any point in time in that they reflect the current contents of short-term memory (a term they use that is analogous to the notion of working memory). In analysing TA verbalisations, then, it is important to take account of the fact that one is dealing with the *products* of cognitive processes rather than with self reported descriptions of processes. What this means is that underlying processes can only be *inferred* on the basis of careful and systematic interpretation of protocol content. One important issue that Ericsson and Simon have addressed relates to whether thinking aloud is a reactive method (i.e. does the TA requirement change or distort the processes being studied). After reviewing a large number of relevant studies Ericsson and Simon conclude that thinking aloud does not affect the structure and course of processing although it may slightly decrease speed of performance. However, if subjects are instructed to concurrently verbalise only selected types of heeded information or information that would not usually be heeded, then normal processes are more likely to be changed. In our study we were clearly careful to avoid recourse to these latter sorts of verbalisation requirement. Another issue that Ericsson and Simon have addressed concerns the completeness of verbal reports. They argue that in concurrent verbalisation the main omissions concern various intermediate products of recognition processes, though the final product will enter short-term memory and may therefore be reportable.

In our second study the actual problem that we used was intended to simulate a real-world design task and was therefore couched in terms of an informal *design specification.* The specification expressed a requirement to design an integrated circuit in an image processing application. Certain of the circuit's functional requirements were detailed as mathematical formulae while other functional requirements necessitated that the engineer spend time in actually developing appropriate mathematical formulae. Clearly then, the problem was selected to address the nature of the design strategies employed in dealing with problems at highly abstract levels of the design hierarchy. All the engineers who participated in this investigation had expertise in working on integrated circuit design problems at these higher conceptual levels.

The verbal protocols acquired in the study provided the main starting point for various analyses while the visual protocols of the engineers' pen-and-paper work (e.g. sketching and note making) formed an important supplementary source of data. Transcribed protocols were each segmented into a sequence of verbal *chunks* using a detailed taxonomy of design behaviours that had been derived from a prior cursory analysis of protocol content. Whilst analysis of the annotated protocols at the fine grained level of minute-by-minute behavioural transitions has yet to be undertaken, a more gross analysis of higher level activity patterns has confirmed some of the conclusions of the previous study. For example, subjects' design strategies were again leading to top-down decompositions of functional modules through levels of increasing concreteness. In the present case, however, the expansion of modules was generally seen to occur in a way that catered for their functional interdependency at each level, that is, a set of modules at one level was developed to a more concrete level in a sequential breadth-first manner.

The pursuit of substantially alternative high-level design concepts, again appeared not to be a pervasive aspect of these engineers' design methods as they tended to become selectively focused on satisfactory solutions rather than investigating other design options that may have proved to be much better than just satisfactory. The gradual iterative improvement of selected design concepts through levels of increasing detail was, however, clearly evident. We realise that the inclination observed for these engineers to pursue a limited number of solution alternatives could have arisen as a consequence of the time pressure on them to derive designs within a restricted period of ninety minutes. Certainly, then, this particular manifestation of a satisficing principle should be interpreted tentatively – although when one considers that real-world design activities are usually heavily constrained by time factors, it may well be another specific instance of a more general tendency.

Analysis of the actual content of the data derived in this study revealed the existence of a variety of errors and inconsistencies in the models that subjects were developing. In considering the likely causes of defective design performance we suggest that cognitive factors which induce the *selective* encoding or processing of information [11] were in all likelihood constraining the exhibition of design competence by these professional subjects. For example, the lack of salience or vividness of problem information could lead to failures to attend to important aspects of the specification. Another possibility is that capacity limitations of working memory might affect performance when concurrent attention to several items of information is required (e.g. whilst manipulating mathematics, whilst investigating competing design options or whilst expanding designs to more detailed levels).

Two subjects in particular showed very clear evidence for the incomplete encoding of presented information in that they formulated only cursory representations of given mathematical expressions. During subsequent design work these subjects were seen to be associating their ill-formed versions of the presented expressions with certain design parameters, when, in reality, these parameters bore no relationship to the actual original expressions whatsoever. Noteworthy is the point that these engineers were the ones who spent the briefest initial time in understanding the input, output and functional requirements of the circuit as specified in the problem statement, and were also least likely to refer back to the specification during later stages of their work.

Many design errors that we have observed in subjects' design work occurred not during the initial encoding of the problem statement but during the subsequent generation, integration and evaluation of solution-oriented information. One good example of such an error was provided by a subject who was engaged in elaborating (as a new functional block diagram) a previous solution model for the overall design problem (which again existed on paper as a block diagram). In this process of solution elaboration the engineer was seen to actually miss out two fundamentally important iterative loops that existed in his original diagram. This clearly resulted in an inaccurate design solution, although this error might well have been spotted by the designer had he attempted to check the consistency of the design expansion which he had undertaken. Evidence for the occurrence of notational inconsistencies was also prevalent in the design data of these professional subjects. They appeared to be particularly prone to (a) using different symbols and abbreviations to refer to the same design parameter (e.g. switching between upper and lower case letters was especially common) and (b) using the same symbol or abbreviation to refer to different design parameters. While the former type of notational inconsistency rarely seemed to cause any problems, the latter type was seen to lead to some fundamental design failures. An example of such an inconsistency was provided by a designer who, within the space of several minutes, used a 'theta' symbol (θ) to refer to two different design parameters. At a later point these two parameters were incorrectly being considered as one and the same, leading to a basic design mistake which was propagated through to lower levels.

When the results of our two empirical studies are considered together, we are impressed by the fact that the design strategies and cognitive limitations observed appear to generalise across very different tasks, design time-scales and levels of expertise. It still remains, however, to confirm the generality of these findings by further research. In this regard we suggest that one potentially valuable study would involve investigating the design activities of professional engineers tackling long-term design problems for real world applications. This kind of longitudinal investigation could provide more insight into the dynamics of the evolution of design solutions in professional settings and could give a clearer indication of the nature of the design strategies adopted by experienced engineers.

Our findings derived from these psychological analyses of engineers' design processes have suggested certain facilities that an intelligent design system could offer. In general, these facilities would be geared toward assisting with the search for alternative and potentially optimal design solutions as well as toward reducing the time of the enginering design process. More specifically they would:

1. Encourage the designer to consider an increased number of initial high-level solution concepts and enable the efficient formulation of alternative versions of each solution concept through levels of increasing design detail,
2. Assist with the choice of competing design solutions, for example, enabling evaluations of solutions to be made on the basis of comparative functional simulations,
3. Superintend the designer's exploratory activity, for example, helping the designer to backtrack if a path proves unpromising (i.e. by providing a record of paths taken together with the current point of exploration) or suggesting worthwhile paths of investigation (i.e. by suggesting design alternatives),
4. Ensure the designer's awareness of design conflicts (e.g. if crucially important constraint requirements have been overlooked when the designer is focusing on a narrow aspect of the overall design solution),

5. Ensure the designer's awareness of inconsistencies in the notation that is being used (e.g. if two different design parameters have been given the same symbolic label).

3. Overview of the PEDA System

3.1. Overall form and function

The form and function of PEDA have been derived from the psychological study with the objective of enhancing the performance of the design engineer without inhibiting his/her natural style of working.

The system is presented to the user as a screen-based drawing board providing mouse-driven direct-manipulation of block diagrams, which are the preferred notation of electronic engineers. Blocks are selected from a palette and represent any level of the functional decomposition of the design from user-specified mathematical functions down to individual electronic components, such as ROM or RAM.

A typical palette appears on the left hand side of Fig. 1 to 3. The largest window in Fig. 1 shows the construction of a diagram to represent the operation of a FIR filter. The popped-up menu displayed on the lowest multiplier shows the options available to the designer for manipulating that block. Fig. 2 shows the diagram completed, with connectivity between components expressed as lines linking communicating parts.

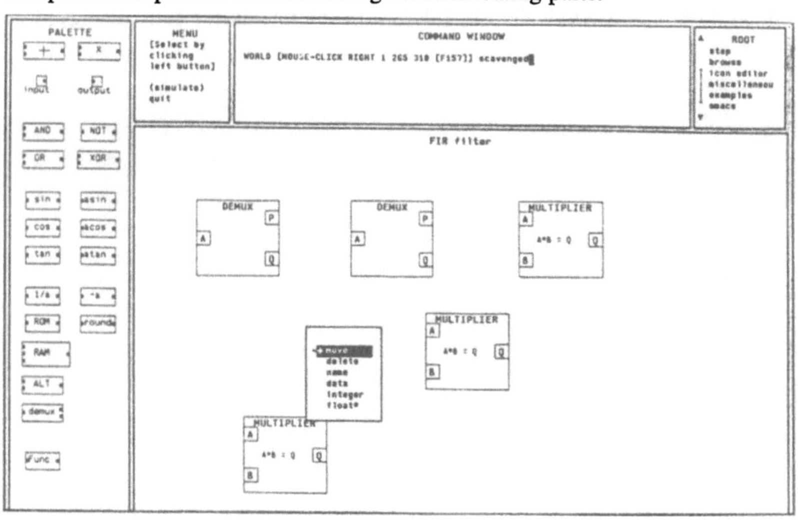

Fig. 1. Diagram construction

The functionality and connectivity represented by the user interface of PEDA is modelled by an underlying functional layer which simulates the actions of individual design objects and operation of the connected whole. The user is able to probe intermediate results and values by

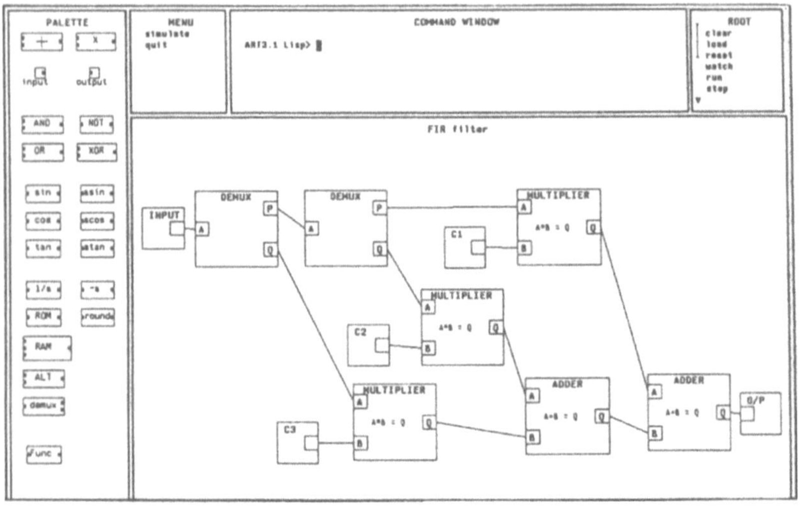

Fig. 2. **Completed diagram**

opening windows at points to evaluate the dynamic behaviour of a design. Fig. 3, for example, shows the effect of sending data through the FIR filter, with intermediate results displayed at different points.

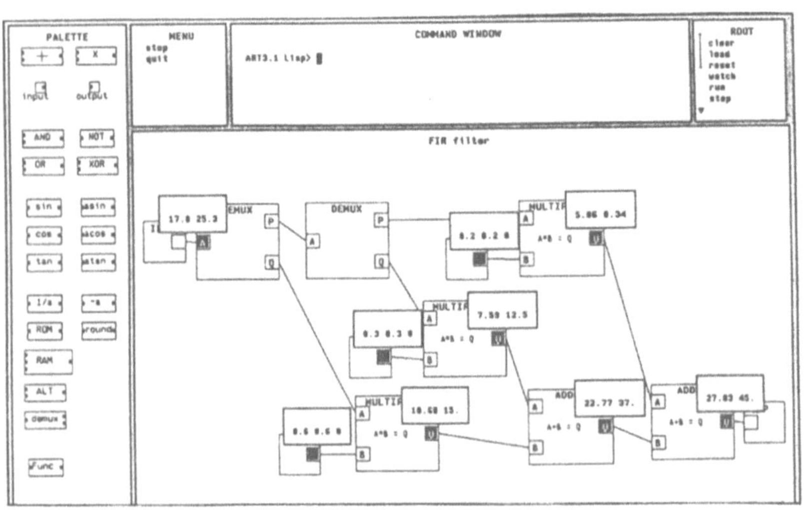

Fig. 3. **Simulation results**

PEDA supports the hierarchical nature of design by permitting several levels of function and the facility to zoom in on high level design objects to manipulate their constituent functions or components. For example, the filter shown in Fig. 3 may be expressed as a single block, with the same overall functionality, in a diagram at a higher level of design.

Optimal, rather than just satisfactory, designs are encouraged by the facility to explore, develop and store many alternative designs in parallel as inter-related versions. The PEDA also has the ability to monitor the evolution of such versions and provide advice on the actual process of design, for example reminders of incomplete activities, a record of crucial assumptions and constraints, and a note of instances when a particular solution has already been explored elsewhere. The system also facilitates backtracking through the design process by means of the logging of design activity.

Knowledge-based functions will eventually enable the validation and optimisation of designs using a data-base of component characteristics in conjunction with a rule-base of electronic design heuristics.

3.2. Implementation system

PEDA is implemented on a Sun 3/60 workstation using ART. The main components of the ART language are facts, schemata, actions, rules and viewpoints. The ART compiler maps the language into LISP and integrates it into the LISP environment. Facts are separate pieces of declarative knowledge. Schemata are used to organise knowledge about objects that are related to each other. A schema is a collection of facts that represent an object or a class of objects that share certain properties. Actions provide a means of attaching procedural code to objects (i.e. schemata or their attribute slots) so that elements of a design can be modelled in terms of their characteristics and behaviour in an object-oriented style. Rules may use forward or backward chaining (i.e. from facts or from goals). There is a powerful pattern-matching language to describe the data that drives the procedures. ART Viewpoints provide a means of modelling hypothetical alternatives and situations that change dynamically with time. It is recognised as one of ART's most sophisticated and powerful features and is particularly useful for the manipulation of related but distinct versions of the users' designs.

This feature of ART would be still more useful if it allowed alternative versions to be stored within. As it stands, additional schemata have to be created to enable versions of designs to be stored for future reference.

4. Functional Aspects of the PEDA System

4.1. Design simulation

The simulator provides the basic underlying mechanism by which data is moved around a block diagram. A block diagram consists of a series of hierarchically structured blocks which have been interconnected in such a way to provide a specified function. Within the block diagram data exists as objects called packets which move between the blocks via links between them. When the functional input requirements to a block are satisfied, the functionality of the block is applied to the input data packets, destroying them in the process. After evaluation, new packets are created at the block's outputs as necessary. Block functionality is expressed as a set of mathematical equations, as this affords a greater deal of flexibility than implementations using a fixed level of functionality, such as logic gates. A library of common low level functions is available – however the important distinction here is that blocks need not be built up from these primitives. An example would be the generation of a block with the function $Q = \sin(A*A) + \cos(B*B)$. The primitive function approach would involve creating an equivalent block from a collection of primitives. The mathematical equation approach, however, has sufficient flexibility to use the function as it is.

The simulator architecture is based on a data flow model and as such there is no explicit representation of time. This choice is in accord with the results of the psychology study, indicating that exact timing considerations are not a necessary part of the early design process, as the engineer derives form and function from the project specification. This type of approach has two main advantages, firstly time information can be treated the same as any other data and fed into a block diagram, and secondly the model maps very well onto an implementation scheme using production rules in ART.

4.2. Design process record and versions

The psychological studies suggest agreement with the general consensus that designs evolve along a number of dimensions. Walker and Thomas [21] describe such a model for design representation in that a design can be viewed as being hierarchically structured with the greatest degree of abstraction at the top. This links in with the fact that in electronic design, engineers have been observed to describe the functionality of a design in a top-down manner. As they descend into the hierarchy the description at each level becomes more specific until, at the bottom level, actual components are specified. Walker and Thomas subdivide their representation into three subdomains – Behavioral, Structural and Physical – and categorise each element of the design into its respective area. The psychological studies, however, indicate that a simple representation containing both functional and constraint requirements together with the actual block attributes at the block diagram level is quite adequate for design representation in the early stages of the design process. Within this context a functional requirement is the most general statement of what a block must do, for example, "an adder must add". Constraint requirements limit the range of possible solutions, for example, "the adder must be fast", whilst attributes describe what a block can actually do, for example, "the maximum propagation delay for this adder is 100ns". Within the PEDA a block diagram can be considered a hybrid structure consisting of elements from all three subdomains – but at a relatively high level in their hierarchy. Blocks can then be represented as schemata with slots for attributes and constraint requirements, with attributes being generally inherited along class lines, and constraint requirements propagated down from higher levels in the block hierarchy (for an example see Fig. 4).

The psychological studies have shown that engineers tend to move between subproblems in a hierarchy of problem spaces which transcend abstraction levels. Within these levels the principle of *separation of concerns* is applied to a limited degree. In its pure form this implies that notation and parameter definitions within a level relate specifically to design considerations within that level and rarely between other level. In actuality propagation of constraint requirements does occur between levels, but the information tends to be symbolic and quite abstract in nature, such as, "the required chip area for the function square root is large".

```
(defschema design-1
  (instance-of world) ; a design alternative
  (must-have (speed fast)) ; stated constraint requirement
  (must-have (chip-area large)) ; stated constraint requirement
  (must-have (power medium)) ; stated constraint requirement
  (must-have (design-time medium)) ; stated constraint requirement
  (design-time low) ; attribute inferred from: integer-multiplier-blocks
  (chip-area large) ; attribute inferred from: integer-multiplier-blocks
  (speed fast) ; attribute inferred from: integer-multiplier-blocks
  (power medium) ; attribute inferred from: integer-multiplier-blocks
  (contains-blocks mult-block-1 ... mult-block-3) ; contains 3 multiplier blocks
)
(defschema integer-multiplier-block-1
  (contains-blocks mult-1 ... mult-16) ; contains 16 multipliers
  (result-data-type integer) ; attribute inherited from: 8-bit-multiplier
  (bits 32)
  (design-time low) ; attribute inferred from: 8-bit-multipliers
  (chip-area large) ; attribute inferred from: 8-bit-multipliers
  (speed fast) ; attribute inferred from: 8-bit-multipliers
  (power medium) ; attribute inferred from: 8-bit-multipliers
)
(defschema 8-bit-multiplier
  (instance-of process-x-block)
  (instance-of integer-multiplier)
  (result-data-type integer) ; attribute inherited from: integer-multiplier
  (function ((setq Q (* A B)))) ; attribute inherited from: integer-multiplier
  (bits 8) ; stated attribute
  (design-time very-low) ; attribute inherited from: process-x-block
  (chip-area medium) ; stated attribute
  (speed fast) ; stated attribute
  (power low) ; stated attribute
  (has-instances mult-1 ... mult-16)
)
(defschema integer-multiplier
  (result-data-type integer)
  (instance-of multiplier)
  (function ((setq Q (* A B)))) ; attribute inherited from: multiplier
)
(defschema process-x-block
  (design-time very-low) ; a standard cell
  (technology process-x)
)
(defschema multiplier
  (function ((setq Q (* A B))))
)
```

Fig. 4. Example of a block hierarchy showing blocks represented as schemata

Additional evidence has indicated that the pursuing of alternatives within a design is in general suboptimal. When exploring alternatives engineers tend to lock onto one *satisficing* solution, instead of adequately pursuing a range of reasonable alternatives – presumably this is due to time constraints or experiential biases. PEDA 9upports the switching of attention between alternatives by maintaining several concurrent versions of any part or all of the design. It is hoped that this environment will partly achieve the desired goal of increasing the number of alternatives that are considered in the design process. Additionally the versional structure within the assistant is used to provide the skeleton upon which the tool builds up a model of the user design process. This is in accord with Mostow [16] in that the design can be used as a basis of the design process model. At a simple level the versional structure is used to hold the individual block diagrams that the engineer has been working on. At a higher level additional information is stored regarding the engineering constraint requirements, functional requirements and assumptions used in the formation of a particular version, together with a history of events. As a design evolves multiple versions of the design will be created containing decisions and criteria which may lead to different final designs. These designs will be interrelated, sharing many hierarchical subcomponents and possibly whole levels. In this way the evolution of the design may be regarded as one of the fundamental dimensions of the design process. A versional structure is outlined in Fig. 5.

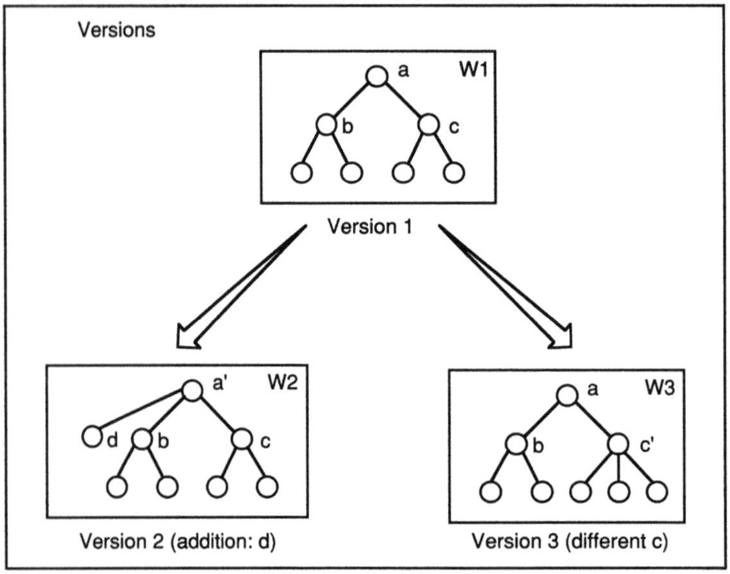

Fig. 5. Outline of a versional structure depicting interrelated designs

4.3. Monitoring

A monitor module within the PEDA is used to convert the interaction between the user and the system into a design trace. This is essentially a record of all the activity since the working session started and exists primarily to prevent the loss of information due to changes in the design. The trace is converted into a design history using an expert system approach to identify and abstract only those parts of the design trace that are considered important to the decision process.

4.4. Design heuristics

Again an expert system approach has been adopted – here to offer advice upon design choices. Rules are derived from both an examination of the literature, especially in the area of decision making in electronic design and a study of designs done within the group, with possible additions suggested by the results of the psychological study. The current heuristics set involve various costing algorithms in design that allow comparative optimisation and validation of designs. These heuristics address simple concepts in both symbolic and actual terms on constraint requirements such as power, chip area and rough speed estimates (see Fig. 6 for an example on the application of design heuristics in the evolution of a hardware design version from a customer specification).

Fig. 6(a) shows what might be a typical mathematical expression declared as part of the customer specification. This would be placed as seen into the system and simulated using an algebraic manipulation package integrated into PEDA. The design engineer developing the project may be specifically exploring a digital implementation and be concerned with the minimal accuracy the user can tolerate. To this end Fig. 6(b) might be a discrete time version of Fig. 6(a). Currently to evaluate different versions of this, PEDA supports arbitrary accuracy functional blocks. Thus mapping Fig. 6(b) to a hardware version in Fig. 6(c) allows the engineer to verify his developments by simulation and subsequent comparison to Fig. 6(a). By selecting different simulation accuracies for each block in Fig. 6(c), several versions of the functional block may be created. For example, using multipliers with 8 bit integer accuracy and 24 bit adder/accumulator accuracy may be a version. There may be many such versions, utilising all possible combinations of Fig. 6(d). The accompanying engineering constraints are available instantly for the whole functional block, for each version, allowing comparisons to be made quickly by the engineer.

Ultimately PEDA will allow similar rules of thumb to be made directly on mathematical expressions. The data in Fig. 6(d) will ultimately be derived via heuristics from real circuit performance characteristics. At the moment PEDA uses a small test set of data in the area of VLSI design – although in reality the standard libraries would apply to a range of specific technologies and include, for example, data on 1.5um CMOS processing. In outline the heuristics are used to compare constraint requirements at a level with a corresponding collective attribute derived from the attributes of the blocks at that level and below it in the block hierarchy. Each versional structure has a corollary in a created ART viewpoint or world which is used by the heuristics to reason about that specific version, with merged viewpoints catering for cross versional inferences. Additional rules act upon the design history and will be able to offer the engineer the reasons for a particular design decision, including the decision path up from the current to the base design. Later, PEDA will offer advice regarding the design strategy adopted by the engineer by comparing the user's design process with an idealised model derived from the psychological study.

$$f_{(x)} = \int_{t=0}^{20ms} x_{(t)} y_{(t)} \, dt$$

Constraint ?slots
Speed over 10 MHz
Design time under 12 weeks

(a)

$$f_{(x)} = \frac{1}{N} \sum_{n=0}^{n=2} x_{(n)} y_{(n)}$$

(b)

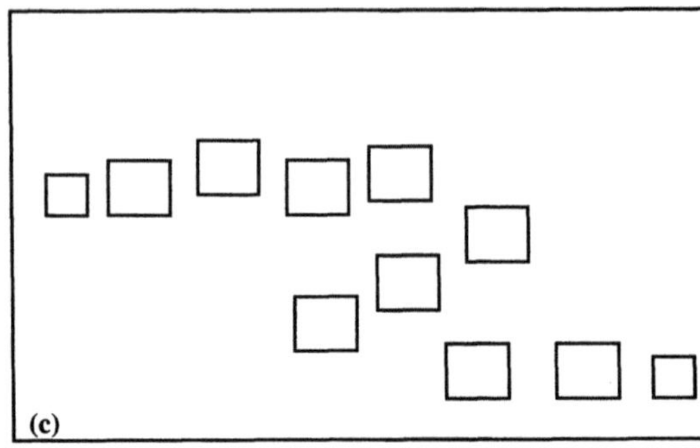

(c)

	Adder				Multiplier			
Size (bits)	8	16	24	32	8	16	24	32
Power	low	low	low	low	low	medium	medium	medium
Design time	u-low	v-low	v-low	v-low	u-low	v-low	low	low
Speed	v-fast	fast	fast	fast	fast	medium	medium	medium

(d)

Fig. 6. Application of design heuristics

5. The User Interface

5.1. Underlying principles

The design of the user interface is based not only on the results of the associated psychological study, but also on psychological work beyond the project. The objectives are to reduce the cognitive load borne by the user and to enhance the human talent being brought to the task of designing. The approach takes account of the fact that humans and machines have complementary strengths and weaknesses; for example, humans are creative but forgetful. A well-designed user interface is particularly important in a co-operative system, since successful co-operation between the user and the machine depends on good communication. The discussion of the user interface will be divided into three parts: the basic interface to the simulator, the representation of knowledge about the instantiated design, and the representation of the design process.

PEDA has a direct manipulation interface in order to reduce the gulf between the user and the system during execution and evaluation, and to give the user a feeling of direct engagement [12]. The system naturally uses a drawing board, rather than the more familiar desktop, metaphor. The user creates his diagram using the mouse to select objects from a palette. Thereafter, a pop-up menu associated with a particular instantiated object presents the user with the choice of relevant, valid functions that can be performed on it. An example is the identification of those ports that the user wishes to probe during a simulation. Figures 1 to 3 illustrate three stages in the use of an early version of the system to simulate a FIR filter, namely: diagram construction (showing a pull-down menu), the completed diagram, and the results of a simulation.

The operation of the interface, described in §3.1, is designed to reinforce the users ability to hold a consistent conceptual model across all of the system's prototype will investigate the success of these and alternative mechanisms.

In addition to the basic facilities of diagram construction and simulation display, the user interface is concerned with the representation of more abstract concepts, namely: design knowledge and the evolution of the design.

5.2. Communication of design knowledge to the user

Design knowledge embedded in the system (see 4.3) enables it to perform heuristics such as optimisation and validation, which must be represented in a meaningful way. The requirement is somewhat analogous to the provision of explanations by expert systems, which are notoriously hard to achieve satisfactorily [3]. A highly graphical approach can be more appropriate in an ICAD system. For example, tradeoff criteria, such as size versus speed, naturally lend themselves to a graphical rather than a textual display.

Alternative methods of representing and explaining design knowledge and advice will also be explored with a number of prototyping studies.

5.3. Communication of the evolution of the design process to the user

The representation of the evolution of the design process is an even more difficult problem, requiring the interface to display an abstract slice through a concrete process, as the system attempts to derive a goal-orientated explanation of the progress of an actual design session.

A simple sequential trace of the session may in some circumstances be useful, for example as an aid to memory. However, the designer is also likely to want other views such as a representation of the "main road" from goals to final design, uncluttered by side-turnings, diversions and dead-ends.

Theory suggests that a pictorial representation of the version tree, showing the relationships between the different versions of the design that the user has explored, will be particularly useful if it highlights such things as constraints, assumptions, conflicts and changes, and thus enables the user to keep track of what he/she is trying to do.

Techniques similar to those used for program visualisation [17] will be evaluated in the prototype.

6. Conclusions

The system described in this paper is distinguished by its basis in a sound psychological study of the design processes used by professional engineers. The psychological work has identified weakness inherent in design practice – in particular, failures to consider sufficient alternative versions to provide optimal design choices – which the system is being designed to correct. In addition, the system addresses the observed need for effective simulation as part of the design process at all levels of the hierarchy and demonstrates the advantages of a direct manipulation interface in this type of application.

References

1. B. Adelson and E. Soloway, ''A Model of Software Design,'' *International Journal of Intelligent Systems*, no. 1, pp. 195-213, John Wiley & sons, 1986.
2. M. Asimow, *Introduction to Design,* Prentice-Hall, New York, 1962.
3. D.C. Berry and D.E. Broadbent, ''Expert Systems and the Man-Machine Interface,'' *Expert Systems*, vol. 4, no. 1, pp. 18-27, 1987.
4. R. Byrne, ''Planning Meals: Problem-Solving on a Real Database,'' *Cognition*, no. 5, pp. 287-332, 1977.
5. R. Byrne, ''Protocol Analysis in Problem Solving,'' in *Thinking and Reasoning: Psychological Approaches*, ed. J.St.B.T. Evans, Routledge & Kegan Paul, London, 1983.
6. A. Dillon and M. Sweeney, ''The Application of Cognitive Psychology to CAD,'' in *People and Computers IV: Proceedings of HCI '88*, ed. D.M. Jones & R. Winder, Cambridge University Press, Cambridge, 1988.
7. C.M. Eastman, ''Problem Solving Strategies in Design,'' in *Proceedings of the 1st Annual Environmental Design Research Conference*, pp. 138-152, 1969.
8. C.M. Eastman, ''Cognitive Processes and Ill-Defined Problems: A Case Study from Design,'' in *Proceedings of the 1st Joint International Conference on Artificial Intelligence*, pp. 19-31, 1969.
9. K.A. Ericsson and H.A. Simon, ''Verbal Reports as Data,'' *Psychological Review*, no. 87, pp. 215-251, 1980.
10. K.A. Ericsson and H.A. Simon, *Protocol Analysis: Verbal Reports as Data,* MIT Press, Cambridge, Mass, 1984.
11. J.St.B.T. Evans, *Bias in Human Reasoning: Causes and Consequences,* Lawrence Erlbaum Associates, Brighton, 1989.

12. E.L. Hutchins, J.D. Hollan, and D.A. Norman, ''Direct Manipulation Interfaces,'' *Human-Computer Interaction*, vol. 1, pp. 311-338, 1985.

13. R. Jeffries, A.A. Turner, P.G. Polson, and M.E. Atwood, ''The Processes Involved in Designing Software,'' in *Cognitive Skills and Their Acquisition*, ed. J.R. Anderson, Lawrence Erlbaum Associates, Hillsdale, 1981.

14. J.C. Jones, *Design Methods: Seeds of Human Futures (Second edition),* Wiley, New York, 1980.

15. J. de Kleer and J.S. Brown, ''Assumptions and Ambiguities in Mechanistic Mental Models,'' in *Mental Models*, ed. D. Gentner and A.L Stevens, Lawrence Erlbaum Associates, Hillsdale, 1983.

16. J. Mostow, ''Toward Better Models of the Design Process,'' *AI magazine*, vol. 6, no. 1, pp. 44-57, 1985.

17. B.A. Myers, ''Visual Programming, Programming by Example, and Program Visualisation: A Taxonomy,'' *Proceedings of ACM/SIGCHI*, pp. 59-66, 1986.

18. A. Newell and H.A. Simon, *Human Problem Solving,* Prentice-Hall, Englewood Cliffs, 1972.

19. G. Pahl and W. Beitz, *Engineering Design,* Springer-Verlag, Berlin, 1984.

20. H.A. Simon, *The Sciences of the Artificial,* MIT Press, Cambridge, 1969.

21. R.A. Walker and D. Thomas, ''A Model of Design Representation and Synthesis,'' *Proceedings of the 22nd Design Automation Conference*, 1985.

22. A. Whitefield, ''A Model of the Engineering Design Process Derived from HEARSAY-II,'' in *Human-Computer Interaction*, ed. B. Shackel, North-Holland, Amsterdam, 1985.

3

A Component Class for Design Objects

Jan Rogier[†]

Department of Interactive Systems
Centre for Mathematics and Computer Science (CWI)
Kruislaan 413, 1098 SJ Amsterdam, The Netherlands

Abstract: This paper discusses an architectural design support system. The system is based on a strategy of incremental spatial decomposition of a design object. It provides mechanisms for the application of strategies such as functional decomposition and re-use of design object knowledge. First we discuss two basic strategies used in architectural design, viz. decomposition and application of prototypes. We then discuss the architecture of a design system that supports both strategies. By means of an example, we show how the system actually supports an architectural design process and how the system is used to communicate with external, autonomous design support expert systems. Such systems provide special facilities for drawing, calculating and other tasks performed during a design process, and are interfaced by means of our design system. The last part of the paper contains an evaluation of our work. We propose topics for further research for our project: the IIICAD project. The aim of our system is to provide a sound basis for the tools developed in the IIICAD project and at the same time to be an interface to already existing tools like expert systems for FEM modelers, drawing machines etc.
Keywords: Intelligent CAD, Knowledge Representation, Design Strategy, Object Oriented Programming.

1. Introduction

One of the main problems in knowledge based design support is the representation scheme of a design object description. Such a representation scheme must fulfill a dual purpose. In the first place, it is an integrated description of the design object, constructed during the course of the design. In the second place, it is used in the process of re-using design object knowledge in other designs. Only a representation scheme that can be used for both purposes, forms a strong basis for the construction of a knowledge based design support system.

Both intentions of the above mentioned design object representation scheme lead to different requirements. The identification of these requirements is based on the analysis of

[†] Also at: Institute of Applied Computer Science (ITI), Netherlands Organization for Applied Scientific Research (TNO), Schoemakerstraat 97, 2628 VK Delft, The Netherlands.

design. In design we identify two basic strategies. A first strategy is incremental decomposition of the design object. It supports the evolution of a design object description from global to detailed, using stepwise refinement [9]. We call every increment in such a decomposition a *design step.* A second strategy is the application of existing design object descriptions to support a design step. Such existing design object descriptions are used as *prototypes* for new designs.

With the above introduced decomposition strategy we capture both the design process itself and the result of a design process. The result is a hierarchical design object description. There are several methodologies for object decomposition. We make a distinction between *functional* and *spatial* decomposition. Functional decomposition uses the functional specification of a design problem. It decomposes a design problem into comprehensible sub-problems and accordingly it provides a functional decomposition of a design object [1]. Spatial decomposition aims at building the structure of a design object. With spatial decomposition, a design object is described as an assembly of spaces with sub-spaces.

The application of existing design object descriptions as prototypes, supports the actual declaration of a design object. A prototype provides the user with a previously built spatial decomposition structure, which serves as a template for further declaration. The user can change this template in order to make it fit to a functional specification. A prototype contains in general more information than supplied by a given functional specification for the design object description. Therefore, a prototype can be used to extent the functional specifications.

In this paper we propose a design supporting system based on spatial decomposition of a design object. In the next section we introduce and elaborate the concept of spatial decomposition and the application of existing design object descriptions as prototypes. In §3 the system and its functionality is described. §4 contains an example that illustrates the application of the system. The last section evaluates the system and proposes topics for further research.

2. Architectural Design Strategies

2.1. Functional and spatial decomposition

An architect's design problem is to assign both space to function and function to space. In both cases he applies a strategy based on *decomposition.* We define *functional* decomposition as assignment of space to function, and *spatial* decomposition as an assignment of function to space. A special case of spatial decomposition is called *material* decomposition, which is a known strategy in assembly based design [2].

Functional decomposition results in a functional specification of different parts of a design object model. A functional decomposition is associated with a specific field of interest. The concept of functional decomposition for architectural design is discussed in detail by Alexander [1]. He solves a design problem by sub-dividing it into different sub-problems, corresponding to different sub-functions. Each of these problems is tackled by an independent sub-system. The design solution is found by integrating the results from each of the sub-systems. Straight on application of a strategy based on functional decomposition, results in a number of aspect models of a design object. These aspect models exist independently from each other. The major disadvantage of the application of functional decomposition, is that it generates a problem of integrating different aspect models into a single, coherent design object model.

Spatial decomposition describes an object as an assembly of spatial parts, which may have sub-parts. Spatial decomposition is devoted to the design object solution. Spatial decomposition combines function with space in a sense that any given (sub-)object is characterized by the fact that the space it occupies is devoted to one function. Such spaces are disjunct, but may overlap resulting in sub spaces resulting in more than one function. An example spatial decomposition is material decomposition. Material decomposition only considers spaces filled with a material.

The integration of functional and spatial decomposition in a design supporting system, and the possibility to incorporate material decomposition in this concept, is based on the next strategy.

- A space is devoted to one complex functions. Aspect models can be produced as required by internal and external applications. These aspect models are reductions of the spatial decomposition which still have the property that the relevant functions are uniquely determined through spatial references. Aspect models therefore are subsets of a model based on spatial decomposition.
- A space is made from one composite material. This composite material has unique properties. Further decomposition of a space into its sub-spaces causes a more detailed material specification of the design object.

We elaborate this strategy as follows: With functional decomposition, as proposed by Alexander [1], for each function, a sub-system or sub-model of the design object is generated. Each function poses constraints on spatial parts or sets of spatial parts. The same spatial part is referred to by different aspect models. For example, a 'wall' of a 'building' will have both a function in the construction of that building and a function of being a shelter for the inside of the building from the outside. All functions of a part has become one unique composite function, as soon as they are associated to a specific spatial part. A part therefore has one function.

Physically speaking materials are mutually exclusive, therefore material decomposition is a special case of spatial decomposition. Material, in architectural design is defined on a high semantic level. On this level all material is composite material. This composite material has unique properties that are relevant in designing and are not directly to be derived from the different sub-materials. As an example consider the case of a 'brick' under 'water'. In this case there are, logically speaking two different materials in the same location (e.g. 'water' and 'stone'). 'Wet stone', however, has unique properties which cannot directly be derived from the properties from 'water' and 'stone' independently. That kind of unique properties is important in designing. Wet stone is for instance a bad material to use for walls in a sleeping room.

Taking the whole strategy into account, both functional decomposition and material decomposition can be expressed in terms of spatial decomposition. We agree that a strategy based on functional decomposition is a commonly accepted approach for specifying a design object. However we claim that a design object description based on functional decomposition is insufficient when used to support the design process. The reason is that a strategy based on functional decomposition results in a number of independent aspects models which causes a problem of integrating these models into a single design object description.

In our approach functional decomposition is applied using autonomous applications that uses local representations of a design object description. These local representations are derived from the same integrated design object description which is based on spatial decomposition. All autonomous applications contribute to this central design object description by extending the complex functions that are associated with the different spatial components. A strategy of material decomposition is supported by associating a single composite material to each spatial part that belongs to the design object description. This composite material may either be solid or not, which allows 'spaces filled with air' to be part of the decomposition. The description is especially aimed at reuse of design knowledge. As a side-effect it is possible to link our approach with international developments on standardization of product data. This is done by combining our representation with *standard vocabularies*. A standard vocabulary is a set of all primitive entities and operations, used in a specific knowledge domain. E.g. some members of a geometric vocabulary are the primitive entities: *block* and *cylinder* and the operations: *calculate volume* and *calculate the conjunction of two primitives.*.

3. Design Support System

In this section we introduce a design system which supports the design strategy described above. Each component of the design system is explicitly accessible. The system is used as a design support tool rather than as a design expert system. This apprentice like behaviour has a twofold reason. The main purpose of our implementation is to illustrate the applicability of the design strategy, presented in §2. The second reason is that we believe that a system should never be more than a tool in supporting the design process. Instead of hiding information from the user, as is often done in so-called expert systems, we state that the purpose of a design system is to expose information. For more detailed considerations on this approach we refer to the work of Bijl [3].

3.1. System architecture

The design support system, described in this paper, is built from a set of components (Fig. 1) It contains a kernel, a database, a knowledge base and a set of external application modules which are accessed by means of application interfaces. The kernel is used to construct a design object model, based on spatial decomposition. This model contains references to entities that belong to the application interfaces and that are stored in what we call *standard vocabularies*. Standard vocabularies are used to access external applications. Applications are used to support functional decomposition of the design object. Support of prototype based declaration of a design object description is done with the kernel. With this kernel a knowledge base and database are accessed. The database contains previously built models of design objects, the knowledge base contains descriptions of standard components and manufacturing features. A detailed discussion on the system's architecture can be found in [6, 7, 9].

3.2. Object-oriented and logic approach

We use an object oriented approach both for structuring the implementation of the experimental system and for structuring the model of the design object. The implementation of the experimental system is based on an object oriented approach in order to get efficient code.

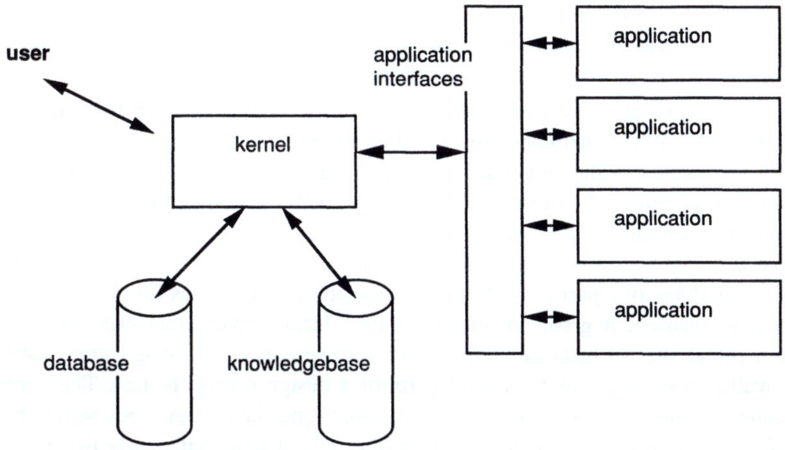

Fig. 1. System's Architecture

Besides it provides a simple and comprehensible structure that is easy to explain. The structure of the model of a design object is based on an object-oriented approach, because this approach best fits on a strategy based on spatial decomposition.

Code efficiency in object-oriented programming is reached by using a fixed hierarchy of classes with subclasses. Each class contains the description of the data template of instances and all functions that may be applied on instances (methods). The hierarchy of classes is an *is-a* hierarchy. Every subclass is a specialization of its superclass. For our system we aim to limitate the number classes that are accessible to the user. In this paper we mainly discuss the class *component* from which part of a model of a design object is an instance. These instances contain references to instances of other classes that are used to define application interfaces.

The model of a design object is also based on an object-oriented approach. In this case however we use different hierarchical approaches. We use an *part-of* hierarchy to describe the decomposition structure of the design object. We use an is-a hierarchical organization of parts to generate prototype descriptions for new parts. This made us implement all parts of a design object as instances of a single class: component. All instances of this class are defined by an extendible number of attributes. Virtual is-a hierarchies of classes are specified by predicates over the set of all instances of the class component. Such a specification allows for a flexible hierarchical organization. We call procedures, that generate virtual is-a hierarchies, *grouping* functions.

3.3. Grouping

A group is a temporal class. A group contains a subset of all instances of the class component that meets a set of predicates. As classes groups are organized in hierarchies. These hierarchies describe is-a relations between classes and sub-classes. We prefer to use the word *group* instead of *class* in order to express the temporal existence of this entity. During their life time however, a group provides the same features as are provided with classes in common object-oriented approaches. Groups are generated by means of grouping procedures that contain a set of rules applied on the contents of all instances of the class component.

Grouping results in the creation of a template. This template contains all attribute names and their types of value [4]. When attributes refer to other instances of the class component or descriptions of relations between values of attributes, these are also included in the template. Generation of a template is controlled by procedures that describe how existing instances should be interpreted. From these candidates the common set of attributes serves as a basic template, while the non-common attributes cause a system controlled interview of the user. As a default mechanism the *group identifier* of a part is used to select appropriate candidates for generating a data template for a new instance. A group identifier is a user declared name that expresses the class to which a user thinks that an instance belongs. To give an example, 'living room', 'bed room', 'kitchen', 'wall', 'door' etc. are all spatial parts of a design object 'house'. They are represented by instances of the class component. 'Living room' and 'bed room' both belong to the group 'room'. What a room is, is described in a procedure. (All instances that have the string "room" as value for the attribute 'group-identifier', belong by default to the group 'room'.) This procedure generates a subset of instances from the class component that is structured in a hierarchy based on their set of attributes (Fig. 2).

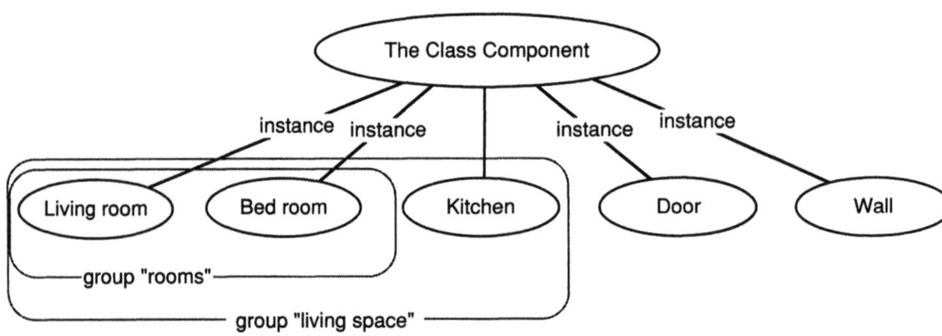

Fig. 2. Classes, Groups and Inheritance

3.4. Unique identification

Each part of a design object is represented with an instance of the class component. This instance can be uniquely identified. Instances of the class component have two identifiers. They have a *name* and they have a *reference* (compare Fig. 4). The name is a string of characters, is local to the instance and does not have to be unique within the system. The reference or pointer is unique in the system, but is only used by the system internally. The user has no direct access to it. Unique identification of an instance is based on identification from context. Every part (always represented by an instance of the class component) is sub-part of another part. On top of this (part-of) hierarchy is the design system itself. Sub-parts of the system are *projects* representing the initial object of a design problem (e.g. 'myhouse' etc.) (Fig. 3). Every other sub-part is part of another part. Although each part has a local name, the system uses a enumeration of names to identify a part by name. This enumeration expresses the part off hierarchy for a specific part, initially starting from the current context. This name may include backward referencing symbols and is as a mechanism comparable with directory identification in UNIX.[1]

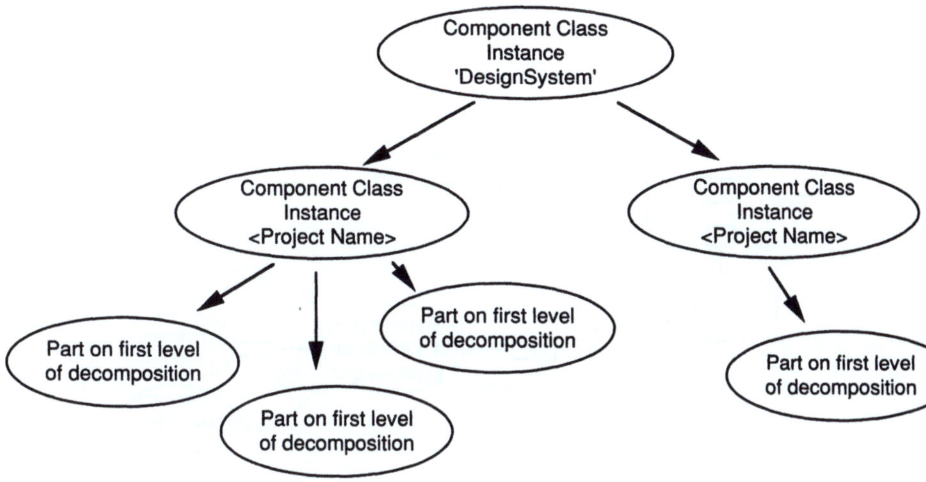

Fig. 3. Has-part hierarchy organization

3.5. Data structure of instances of the class component

An instance of the class component has three variables. It has a name, it has a *group identifier* and it has a *dictionary* with attributes. These attributes serve three different tasks (Fig. 4).

1. They are used to define constant type properties of the part. For example, the attribute 'colour' has a constant value.
2. They are used to describe the decomposition structure of the design object. For example, an attribute 'leg_1' of a 'table' contains a reference to another instance of the class component.
3. They are used to describe relations between different parts or constraints on a part. For example, a reference to an expression that describes the relation between the 'lengths of legs' of a 'table', is a value of an attribute of this table.

Each of the above tasks is denoted by the type of the attribute value. This *type* denotes the attribute (e.g. constant typed attribute, attribute referring to a part, attribute referring to a formula expression).

[1] UNIX is a trademark of AT&T Technologies, Inc.

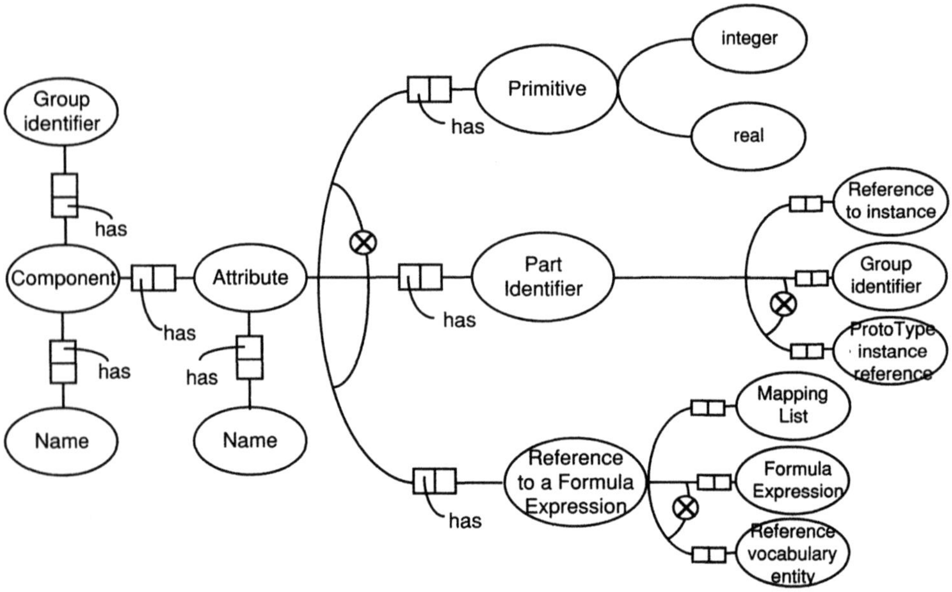

Fig. 4. Data model

3.6. Instance creation and declaration the attribute template

New parts are created as sub-parts of existing parts. This implies a top-down strategy in declaring the structure of the design object. Declaration of the attribute template of a instance can be done in three ways. i) The set of attributes can be declared one by one, ii) references can be made to predefined entities that belong to a standard vocabulary and have a fixed set of attributes, and iii) prototype based declaration strategy can be applied by making copies of either standard components, previously defined objects, or templates of objects that are generated using grouping procedures. Like in ii), creation and declaration of the attribute template are a single action.

The first approach is outrageous and will probably only be used in specific circumstances. The second approach is used to describe aspects of the design object. This approach has a twofold purpose. It is used to detail the model of the design object and it is used to establish a link with external applications that evaluate this model. In the example we elaborate the application of a geometric vocabulary. The third approach provides a template that is generated with a procedure that evaluates the contents of the systems' database or knowledge base. In both cases instance creation and the generation of the attribute template are one and the same action. Standard components belong to the system's knowledge base. A template based on a standard component can not be changed structurally but only for a limited set of attribute values. The system's database contains previously built models of design objects. The contents of this database is evaluated by means of a (selection) procedure. This procedure is an intermediate value of a part-reference attribute. The result of applying this procedure is a attribute template.

When the procedure results in the selection of a single instance, a deep copy of this instance is made including all constant values of attributes. When the procedure results in a set of instances the same procedure has to define a strategy to combine information in a single template.

4. Mathematical Dependencies between Attribute Values

Relations between parts are described by means of (formula) expressions. These expressions describe the (mathematical) relation between constant values of attributes of parts. The description of a relation itself is not part of the design object description, but is stored in an external library. The connection between the design object description and the relation is made by means of a reference and a *mapping list.* The mapping list describes the mapping of attribute-names used in the design object description on attribute-names used in the description of a formula expression. A formula expression has a twofold function. It is used to calculate values of attributes and it is used to verify values of attributes. The specification of a formula expression can be used to guide the detailization of a part. Each of the attributes of an expression has to be mapped on an attribute of the part description.

Formula expressions are stored in libraries. Sets of formulas can be combined into a single entity called primitives. Primitives are also stored in libraries. When these primitives has a twofold functions in a sense that they are used as basics in the communication with external applications, we call these libraries standard vocabularies. Apart from a set of primitives these vocabularies include functions used to access external applications. An example of an standard vocabulary is elaborated in the example.

5. Maintenance of Consistency

Maintenance of consistency is considered on two levels: on the level of the design object description itself, and on the level of the design object description in the context of the system's data base.

5.1. Design object description

Maintenance of the design object description is carried out by means of values of attributes and the evaluation of formula expressions. Each attribute of a design object with a constant value (e.g. 'height'), may have at most three values: a *set* value, a *calculated* value, and a *default* value. The first one is (explicitly) set by the user; the calculated value is the result of executing a formula expression; the default value originates from a prototype for the design object. The set value prevails above the calculated value, the default value is only used for informational purposes but plays no active role in calculation procedures. The consequences of adding and overwriting values are controlled by an user declared procedure, linked to the object description. From each of the attributes of an object is known (is declared) what will happen when its value changes. As a default mechanism every change will cause a warning procedure. We adapted this approach for experimental purposes only. It is subject to further research.

The existence of default values are used to express the difference between re-use of old designs and the use of so-called *standard components.* Copying of an 'old' design object results in a copy where all constant values are used as defaults (for informal query purposes only);

applying a standard component results in an copy where all (or a predefined subset) constant values are interpreted as set values.

As soon as all values used in a formula expression has a value (either set or calculated elsewhere) the expression is used to verify the values of these attributes. If the values does not correspond, based of the evaluation procedure linked to the object, a warning is produced.

5.2. System's knowledge base

Maintenance of consistency on the level of the system's knowledge base is done by restricting the possibilities to refer directly to other object description. As soon as a design object description is used as a prototype for another design object description, it can not be changed any longer. Changes to be made to the prototype cannot be carried out, except by making a new object description that uses the existing one as a prototype. This approach causes problems on conceptual level. In this approach it is not possible to access the properties of number of objects except by incorporating them in a single project. So, for example, changing the 'colour' of a set of 'chairs' is not carried out by changing that property of the object that is used as a prototype, but by declaring that the whole set of chairs that belong to a project and have some common property that is defined on the level of the project; e.g. 'colour of all chairs' (see Fig. 5) The system's decision when to lock a design object description (e.g. prevent further elaboration of the description), is based on the existence of a prototype reference to that object.

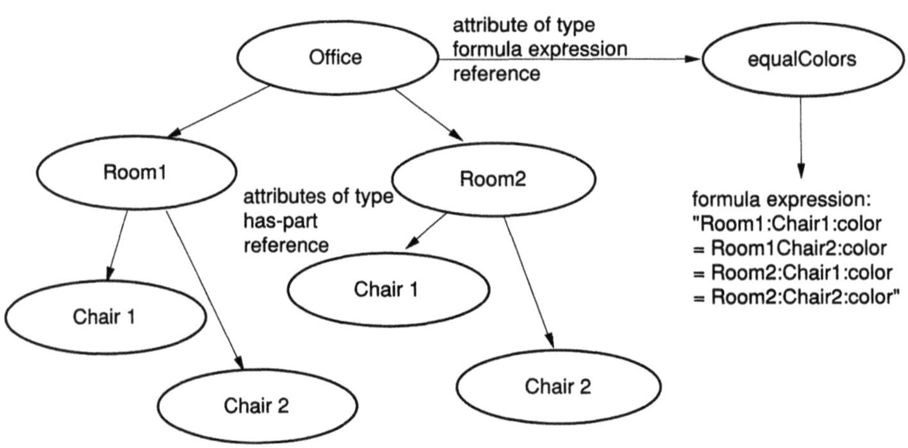

Fig. 5. Relations between different parts

All restrictions together make it possible to store only the new information about an object. All existing information is available for re-use but only (virtual) copies can be used in the context of a new design object description. As a consequence, the first project ever to be done on the system will cost a, relatively, enormous amount of space and specification time. On the longer term, and depending how wide the application field is, each new project will relatively occupy less space and probably less effort.

6. Example

This paragraph contains the description of an example application of the system for architectural design purposes. It describes the design of a simple cottage as people in Holland often build on their allotment. Using an example like this, is based on several considerations. In the first place the example results in an independent object which does not have to fit into a context. In the second place the object is relatively simple. In the third place a cottage is, compared to other buildings, relatively weakly defined which gives the designer a lot of freedom. This makes the example easier to explain. In order to simplify the example to a further extend, we will not focus on generating alternative solutions. The example is meant to illustrate how design can be supported in our approach, not to generate the best cottage ever designed. Another restriction is, that we will concentrate on the description as if it was the first one ever made with our design system. Explanation of the strategy of *prototype application* by means of an example, makes this example unnecessarily complex and will not elaborate this topic.

The example is described in three sub-paragraphs. The first paragraph contains a abstract description of what happened in the design process. It also compares this design process with more complex design processes. The second subparagraph describes the design process in terms of spatial decomposition. The third paragraph focus on a specific design step. In this paragraph is described how an external application will be used to support the process.

6.1. Design process

In our system we apply a design strategy based on spatial decomposition and which contains the following tasks (Fig. 6). The first task is to specify global constraints on the design object. The second task is to translate these constraints into properties of the design object. The third task is to decompose the design object into parts. This task is further sub-divided into several sub-tasks in during which further decomposition takes place. The fourth task is to make links between the decomposition structure and items that belong to libraries of standard components and libraries of *features*. Standard components are the lowest level of decomposition of a design object. Features are descriptions of manufacturing processes which are identified by means of their result. An example of a standard component in our case is a 'door' with a 'door-frame,' an example of a feature is a specific 'wood joint.'

Although subdivision into different tasks insinuates that the process as a whole is sequential, the process is in fact iterative. Which tasks are performed, depends on the amount of information that is actually present at a specific moment. So, for example, cost analysis is only useful as soon as information is present about materials and features, while constructional aspects can only be taken into account, when information is present about the possible location of construction elements. Evaluation of different aspects, in our approach, takes place by means of external applications. These applications internally build a representation of the design object description (e.g. a sub-set of information) and produce information that can be added to the model of the design object. The process is illustrated in Fig. 7.

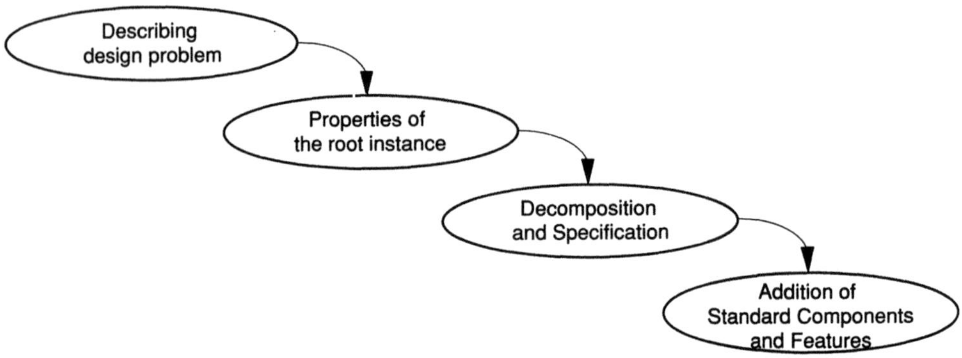

Fig. 6. Design process structure

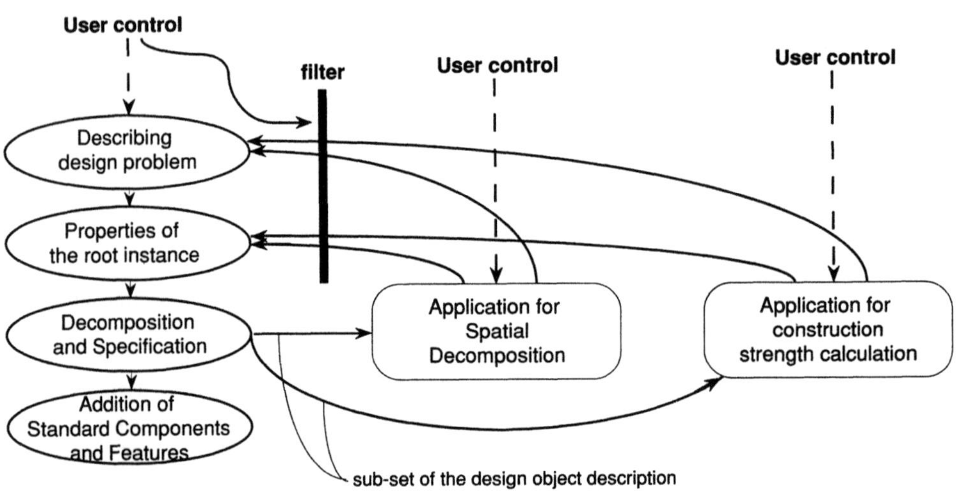

Fig. 7. The role of external applications

6.2. Design object description

For the sake of simplicity, we skip the iteration issue in this approach and describe the process as if it took place in a single sequence.

6.2.1. Task 1. The first task in the design process is taken by collecting some global properties of the design object:

- The most important restriction on cottages is that they may not be used for permanent living.

- They have to be built from 'wood' or another non-permanent material, they may not have permanent sleeping facilities (you may stay in it for a couple of days but not permanently), and they have to be built on about 12.5 square meter (measuring 2.5 by 5 meters) and may be not higher that 2.7 meter. The backgrounds of these limitations are found in the origin of such a cottage, being a dry storage place of garden tools, and a shelter to protect a gardener from rain.
- They may however have a kitchen and a bathroom including a facility like a shower. Although cottages on allotments are not intended for permanent living, they more and more are used as summer houses. So we specify some more requirements based on its function as a summer house.
- There should exist a spatial distinction between what can be called a 'quiet area' and a a 'noisy area,' separated from each other by a 'walking space.'
- Another consideration is that the bathroom and the 'quiet area' should be separated by at least two internal walls.

6.2.2. Task 2. We start design support with initiating a new project called 'De Zonnehoek'. This name identifies a new instance of the class component, the root instance of the new project. Attributes of this object are some global properties; 'height', 'length', 'width', 'material', and some functionality in the sense of constraints on the values of attributes. For representational purposes a geometrical entity is used that belongs to a geometrical vocabulary. This assignment is carried out by means of a *geometrical attribute.* At the same time initial values of all attributes are declared. The declaration procedure has a result illustrated in Fig. 8.

6.2.3. Task 3. The next task in the design process is to decompose the basic object into parts and parts of parts. All parts describe spaces that in a later task will be or be not 'filled' with material. The result of this task is a entity-relationship description of the design object. It contains all relations and some values used as boundary indications of values of attributes. In the process of detailing, the entity relationship model is used as a frame [5]. 'Zonnehoek' contains the non-material 'sitting space' (representing the 'quiet area'), 'walking space' and 'working space' (representing the 'noisy area'). It has four walls, a floor and a roof. The floor plan of the object is illustrated in Fig. 9. Generation of this floor plan lay out, takes place by means of an external application (in our case; the mind of the designer). Computational support of this sub-task will be discussed in the third sub-paragraph.

Also during this task, the model is extended with formula expressions that describe the relation between values of attributes of parts. Relations between values of attributes of different parts or relations between values of attributes of the root object and attributes of parts, are expressed on the level of the assembling part. So the expression "(Zonnehoek:)height = floor:height + walkingSpace:height + roof:height", is specified on the level of the root object 'Zonnehoek' and not on the level of the parts. A part of the decomposition tree is illustrated in Fig. 10.

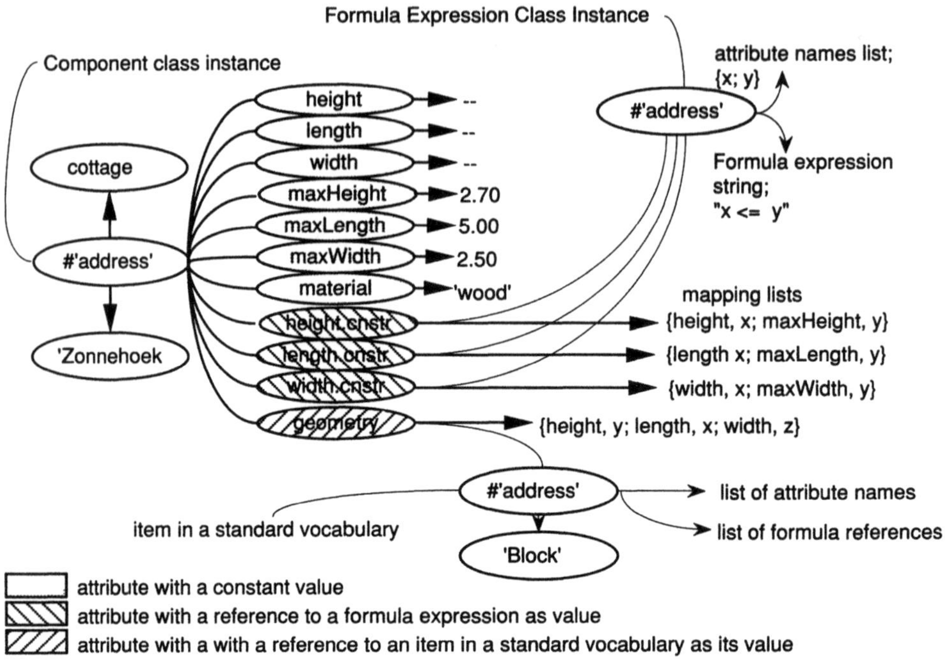

Fig. 8. The data structure in use

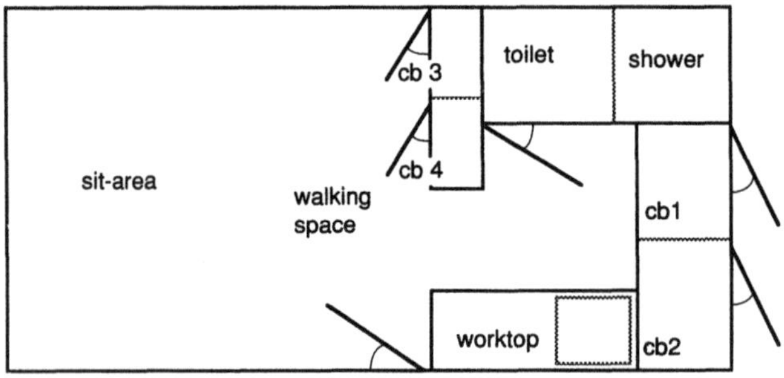

Fig. 9. Floor plan

6.2.4. Task 4. The next task is to 'fill' spaces with material. When a more complex design object was chosen, this would have been the time to use a construction strength evaluation application. In the current description, all places where construction elements can be located are identified and an algorithm is applied to find the optimal location of certain construction elements. In our case (e.g. a cottage on an allotment) we will simply assume that all walls together will be strong enough to carry the weight of the ceiling and will make the total building stable (designer's experience). 'Filling' with material on the one hand takes place by further

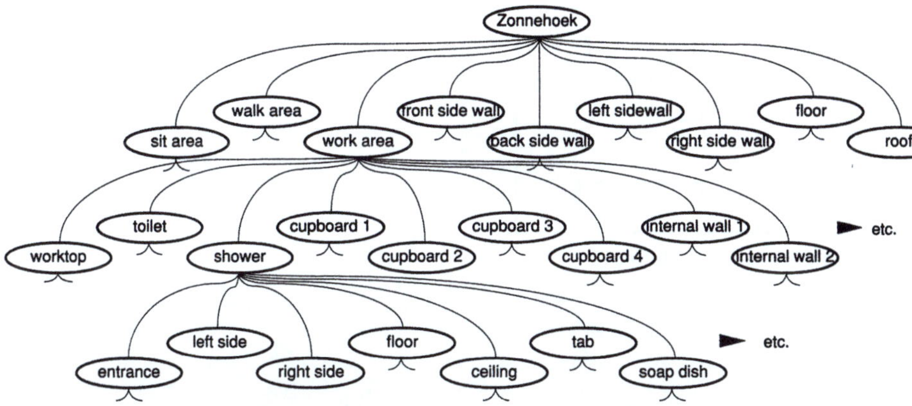

Fig. 10. Part of the has-part hierarchy

detailing the design object description, and on the other hand by choosing solutions from a library with standard elements. In our case we assume that 'doors' and 'windows' may be found in a catalogue and does not have to be designed. On a more detailed level each of the materials for construction, e.g. 'wooden beams' and solutions for 'wood joints' may also be found in a library with standard solutions. Selection of all these standard elements provide our frame with concrete information from bottom-up (e.g. all values of attributes). The last phase in this task therefore integrates a top-down and bottom-up approach in the declaration of values of attributes until the design object description is fully specified. The result of this task is a complete and detailed design object description that can be justified in the context of the consulted external applications (in our case the designer's mind) and be represented geometrically.

6.3. Spatial reasoning

This sub-paragraph gives an impression how spatial decomposition of a design object may be supported by means of an external applications. Although the basics of such a spatial reasoner are implemented, the application itself is not yet operational. The current illustration is therefore hypothetical. The process of spatial decomposition takes place on a design object description as is illustrated in Fig. 11, and results in a description as is illustrated with Fig. 11 (in 3D). The relation between these two descriptions is established, based on a definition that a *space* can be represented as a *space enclosing a surrounding space.* So the original 'Zonnehoek' was geometrically represented by the entity *block.* This block is decomposed into a block with a shell that is itself a space. Decomposition of that shell will result in the construction of a set of (at least) six blocks that enclose the internal block (see Fig. 12). In this case, the relation between the enclosing blocks and some parts of the object 'Zonnehoek' are obvious. All blocks can be used to describe the geometrical properties of parts of 'Zonnehoek' being its four walls, its floor and its roof. The same strategy is applied with spatial sub parts of 'Zonnehoek'. Each of its sub parts is represented by a block that is transformed to a set of blocks.

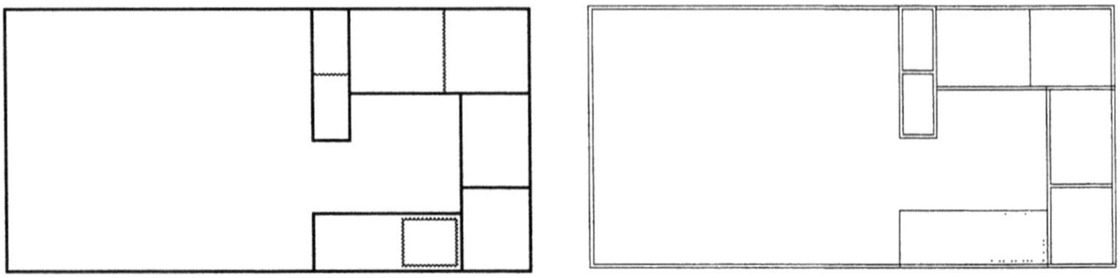

Fig. 11. Design description evolution

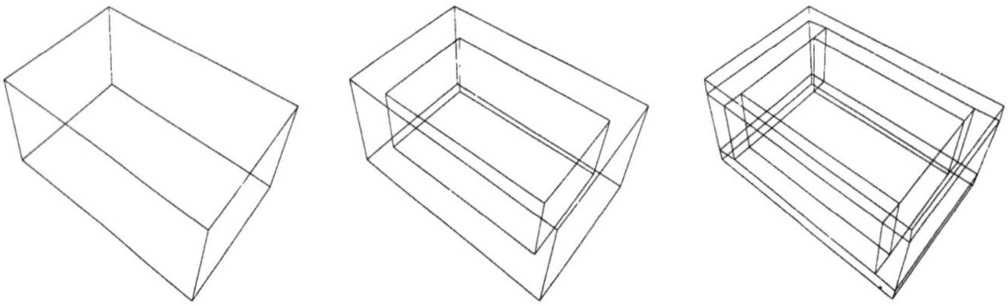

Fig. 12. Materialization of a block

The advantage of the approach is obvious when the relation between representations of spatial parts are described by means of set operations. The seven blocks representing the geometry of the object 'Zonnehoek', have by definition one ore more parts of their surfaces in common with others (in the case of the internal block: all) and do not coincide with each other. The relation between each of these representations with the geometrical attributes of sub parts is described as follows. On the level of the global definition (e.g. one block instead of seven) the space is completely subdivided into non-coinciding block (spaces) that represent the geometry of the sub parts. ('Zonnehoek' is subdivided into 'a quite area', a 'walking space' and a 'noisy area'). On the level of transformation of one block into seven, both horizontally (between sub-parts) as vertically (between a sub-part and a parent part) there may exist overlap (as is illustrated in Fig. 13) but no overlap may take place between an internal space of a sub part and an enclosing space of a parent part (no external wall halfway the interior of a toilet).

This consideration has some consequences: In the first place the user has to provide a description of the relation between the geometrical representation and its transformation. This can take place by means of labels that identify such relations (see Fig. 14 *exterior, interior* and *global).* In the second place, the user has to supply rules concerning overlapping spaces (e.g. ''no overlap may occur on both 'bathroom enclosing space', 'working space area' and 'walking space area','' indicating that the bathroom may not lie against the walking space wall, or anther rule: ''overlap of the spaces enclosing the toilet and the shower results in removal of this space''). These relations can be expressed using set operators: ''conjunction of 'bathroom' & 'walking space' & 'working space' empty'' or ''conjunction of 'toilet' & 'shower' is 'non-material'.'' In the third place optimization criteria should be supplied to the reasoner, saying for

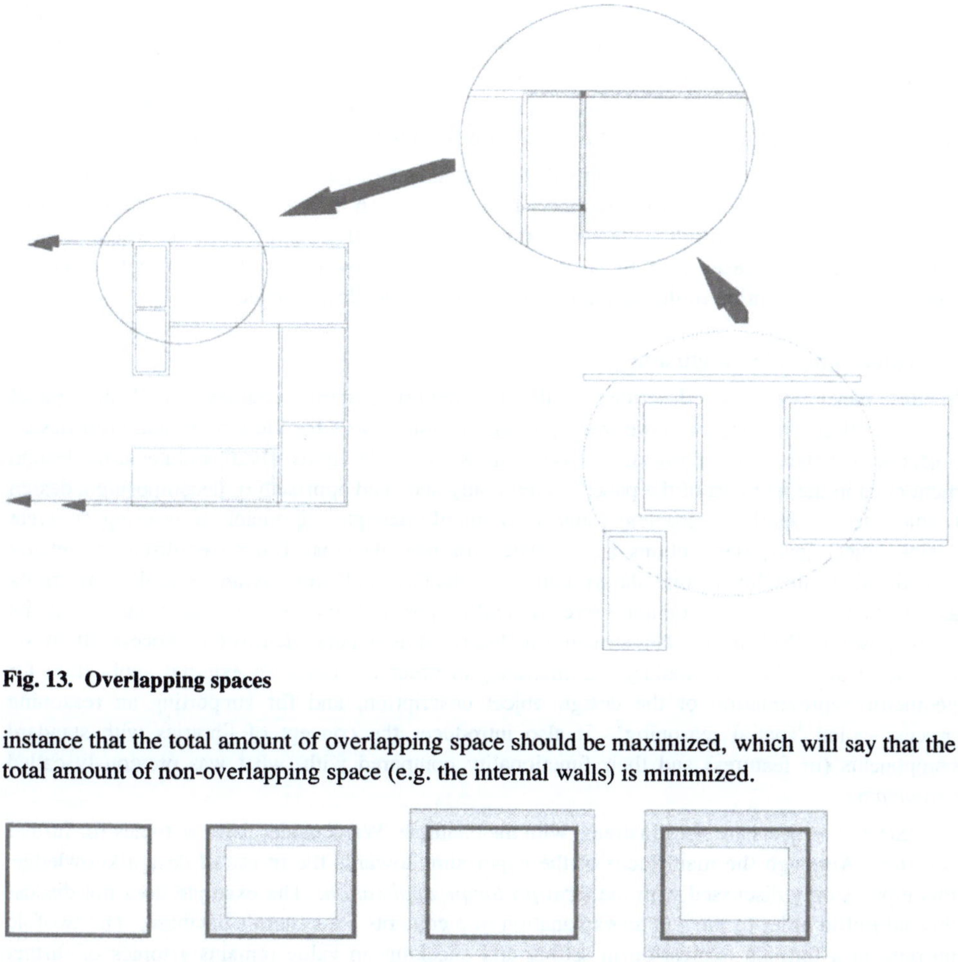

Fig. 13. Overlapping spaces

instance that the total amount of overlapping space should be maximized, which will say that the total amount of non-overlapping space (e.g. the internal walls) is minimized.

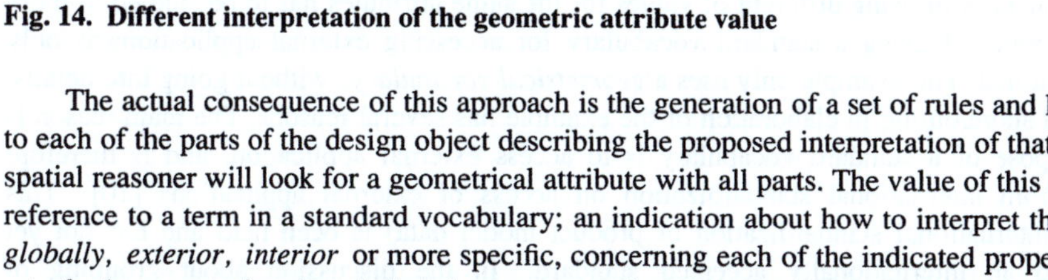

Fig. 14. Different interpretation of the geometric attribute value

The actual consequence of this approach is the generation of a set of rules and labels linked to each of the parts of the design object describing the proposed interpretation of that part. So the spatial reasoner will look for a geometrical attribute with all parts. The value of this attribute is a reference to a term in a standard vocabulary; an indication about how to interpret this term (e.g. *globally, exterior, interior* or more specific, concerning each of the indicated properties); and a set of rules concerning the relations between parts in terms of set operations. Interpretation of the geometric attribute can be based on the existence of upper and lower bound for an attribute value. So existence of 'maxLength' indicates that 'length' should be interpreted as an external measure, 'minLength' indicates that 'length' should be interpreted as an internal measure, while existence of both these attributes indicates that 'length' is a global value. Not that all these attributes, are properties of a space, and not of the object (sub-part) that should be located in that space. So the difference between 'maxLength' and 'minLength' does not indicate the 'thickness' of a wall (to be exact the thickness of two wall together) but a maximal thickness.

As a result and based on concrete values of attributes of parts, different lay outs may result from the application of the spatial reasoner. These results can be manipulated by adding rules or changing values, until a result is reached that is appropriate. The next task in the design process is to 'fill' all these spaces with actual material components which on their turn will provide a more detailed description of the spatial decomposition of the design object.

7. Evaluation and Conclusion

In this paper we have described a design support systems strategy based on spatial decomposition. This approach we think provides a sound basis for the representation of design object knowledge. The approach is based on two considerations about architectural design, mentioned in the first part of the paper: a commonly accepted approach of decomposing a design problem into several sub-problems, and a commonly accepted approach of re-using concrete design object (part) descriptions as templates for new designs. The paper discusses several considerations that lay behind the system's architecture and presents an example concerning architectural design. This example serves several purposes. It focuses on the approach of spatial decomposition. With it, the data structure is described as is generated in this process. It shows the use of a standard vocabulary (geometrical) in order to access an external application for geometric representation of the design object description, and for supporting an reasoning process called 'spatial reasoning'. It also introduces the concept of libraries with standard components (or features) and their functionality compared with, what was previously called *prototyping*.

Several subjects are not illustrated with the example. We consider these as topics for further research. Although the main focus of the paper aims towards the re-use of design knowledge, this topic is only discussed with the term *prototype application*. The example does not discuss this subject in order to prevent an explanation of a contents a a system's database. The twofold purpose of a formula expression in setting and checking an value remains a topics of further research. The current example simply assumes a maintainable model of a design object, but the exact consequences of using different of values for the same attributes has to be studied in more detail. The topic of using a standard vocabulary for accessing external applications is only globally identified. The example only uses a *geometrical vocabulary,* without going into details. This twofold shortcoming in elaboration of the example has several reasons. The main reason is that the purpose of a standard vocabulary is to access external application, and is therefore dependent from international standardization on access of external applications [10]. This discussion (international standardization of product model data) is been held and has not yet resulted into an internationally accepted standard. In the discussion about grouping of comparable part description, we mention procedures that force these (temporal) grouping. Although the purpose of this mechanism is explained, we did not elaborate this topic further. These procedures are based on the description of parts (e.g. their group identifier, their list of attributes names etc.). The default approach is to group all parts with the same group identifier and extract their common attribute (names). The syntax and semantics of more advanced procedures, that will simulate class hierarchies as are presented by 'Smalltalk-like' systems, remain a topic of further research.

With mentioning these topics it looks as if our system needs a considerable amount of further elaboration. Most of these topics however are covered by other products of the research project our system is developed in: the IIICAD project [8]. In this project a design language is made that will cover all the above topics and will be used on top of our system. The aim of our system was to provide a sound basis for the application of such a language.

References

1. Chr. Alexander, *Notes On the Synthesis of Form,* Harvard University Press, Cambridge, MA., 1964.
2. J.-P. Barthes and K. El Dahshan, "Implementing Constraint Propagation in Mechanical CAD Systems," in *Intelligent CAD Systems II: Implementational Issues*, ed. P.J.W. ten Hagen and P.J. Veerkamp, pp. 217-227, Springer Verlag, Berlin, 1989.
3. A. Bijl, "Strategies for CAD," in *Intelligent CAD Systems I: Theoretical and Methodological Aspects*, ed. P. ten Hagen and T. Tomiyama, pp. 2-19, Springer Verlag, Berlin, 1987.
4. H. Lieberman, "Using Prototypical Objects to Implement Shared Behavior in Object Oriented Systems," *OOPSLA'86*, vol. 21, no. 11, pp. 214-223, ACM Sigplan, New York, 1986.
5. M. Minsky, "A Framework for Representing Knowledge," *Readings in Knowledge Representation*, pp. 245-263, Morgan Kaufmann Publishers, Los Altos, 1985.
6. J. Rogier and F. Tolman, "De formulering van een ontwerpparadigma voor bouwkundig ontwerpers," *Ontwikkelingen rond industriele automatisering*, pp. 3-395, SAMSOM, Alphen a/d Rijn, 1989.
7. T. Tomiyama and P.J.W. ten Hagen, "Organization of Design Knowledge in an Intelligent CAD Environment," in *Expert Systems in Computer Aided Design*, ed. J. Gero, Amsterdam, 1987.
8. P.J. Veerkamp, V. Akman, P. Bernus, and P.J.W. ten Hagen, "IDDL: A Language for Intelligent Interactive Integrated CAD Systems," in *Intelligent CAD Systems II: Implementational Issues*, ed. P.J.W. ten Hagen and P.J. Veerkamp, pp. 58-74, Springer Verlag, Berlin, 1989.
9. B. Veth, "An Integrated Data Description Language for Coding Design Knowledge," in *Intelligent CAD Systems I: Theoretical and Methodological Aspects*, ed. P. ten Hagen and T. Tomiyama, pp. 295-313, Springer Verlag, Berlin, 1987.
10. P. Wilson and P.R. Kennicott, *STEP/PDES Testing Draft; St. Louis Edition 4.1.2,* ISO, St. Louis, 1987.

4

Discussion Session: Design Process

Aart Bijl

1. Papers

The opening session of this Workshop received presentations of three papers, under the heading of Design Process. These papers were wide ranging. Jan Treur's presentation on a Logical Framework for Design Processes took us deep into logical abstraction, focusing on formal expressions of requirements and inferencing processes that can generate solutions. A key interest was the formal treatment of partial requirements and the use of unsatisfactory solutions as a prompt to reformulate requirements.

Discussion of this paper prompted some contention among Mac an Airchinnigh and others over the suitability of logic for modelling design behaviour, versus the use of hierarchical object oriented programming. This contention reappeared several times in later discussions, without a firm conclusion. It seemed to rest on different histories and consequent allegiances of participants, reflecting different involvements in system implementation issues and scale of applications. Further deep discussion might have established some convergence at a theoretical level, separated from considerations of implementation in computers.

Phil Culverhouse's presentation on a Psychologically Based Intelligent Design Aid referred to protocol studies and brought us face-to-face with users (designers). It argued for the need to take into account what is going on within users when we devise systems. The stated goal was systems that could monitor user actions, serve as councillors and check for user errors, but stopping short of systems that would know users' domain knowledge or take responsibility for generating design solutions.

During discussion of this paper, Barthès aired questions about the superimposition of programming on top of art, the place of back and forward tracking, addition of design heuristics, interpretation functions, and support for 'what if' strategies. McCullough pointed out that we do not have a taxonomy of design behaviours, and would this need to be biased towards goal seeking, formulating and satisfying? Furthermore, when observing design behaviour are we confident about the difference between 'showing' and 'doing'? Tomiyama referred to other protocol studies, and suggested improved techniques for abstracting protocols by getting two designers to collaborate on the same task, and choosing designers from different fields – they have to say more. Finally, Brower asked what time had been devoted to analysing protocols and referred to techniques for speeding analysis.

Jan Rogier's presentation on the Use of STEP in an Intelligent Design System for Architectural Design focused attention on a specific technique for integrating multiple contributions to designs of single products. This entails mapping various vocabularies of different designers to a single vocabulary (STEP) for purposes of executing interpretations of contributions passed between designers, resting on an assumption that mappings can include sufficient language structure to satisfy each designer's view of a product. The presentation focused on geometry and topology of objects, as typed 3-D regions, which are used in descriptions of products.

Mac an Airchinnigh opened the discussion on this paper with an extravagant discourse on 'wives and kitchens', suggesting different perceptions of space that do not employ aggregations of primitive shapes. Ten Hagen sought to moderate this point by saying that STEP is accepted, can be used to capture the evolution of products, and is intended to help dialogue between designers, not product users. Barthès called for clarification on how the disparate forms of information (e.g. NC code for machining, low level geometry models, etc.) can be encompassed by a single technique and what was meant by history (evolution), does that include errors, and purpose?

2. General Discussion

The moderator offered some general observations on progress that has been made during this series of three ICAD workshops. A shared definition of design remains elusive, but we now exhibit a common motivation (or interest) which recognises issues such as: unexpected input/consistency/completeness; errors/perceptions/intentions; language/representation/purpose. Over the years we have come to be familiar with each other's vocabulary, but we still seem to use many different words for the same meanings – we still speak different languages. Nevertheless, our main achievement, and this is an important achievement, is that we are beginning to understand each other's questions.

The three papers of this session covered a broad range of interest, from formalisms in logic, through design behaviours in persons, to an elaboration of technical implementation issues in machine systems. Logic offers a neat clarity which is appealing, and applications in user domains perhaps inevitably entail confusion. We need to establish robust and general logical treatments that are meaningful to users, designers, and which can survive complexity and scale of applications.

Arbab then reopened the question of how we define design: now we appear to be looking for a definition in terms of top-down refinement, functional blocks, product views. Culverhouse spoke of definition by field, by sets of requirements leading to products, achieved in agreement with product users. Takala included the task of finding requirements. Arbab referred to the generally accepted need for analysis, to help designers understand requirements – but how do designers understand requirements? Culverhouse claimed that designers require adequate databases of domain knowledge. At this point the moderator suggested that we need not go further in seeking to define design – successful development of computers that can aid designers might not be dependent on knowing what designers do, but rather on designing devices which will be useful to designers irrespective of what they might decide to do.

Mac an Airchinnigh aired the issue of time lapse between a designer doing something and later recalling a view, when original intention is lost and renewed explanation is required – and suggested a role here for expert systems. This prompted speculation about random walks through design tree hierarchies, system generated explanations at time intervals, and Desai's critical observation on the usual 'tunnel to goal', leading to his advocacy of breadth first search.

McCullough commented on the danger of design process and product separation, resulting in loss of semantics. Ten Hagen addressed the same point to Treur's presentation, referring to definitions of different views and their eligibility for a system's reasoning processes. Treur said this raised complex control problems, which require research over many years – he claimed that designers' exercise of control leads to mathematically intractable problems, and we need to discover more about how humans know and do things.

Bernus expressed the view that the design situation itself is not our issue, it is implicit in end results and the process is autonomous (within designers). Culverhouse added that protocol analysis of designers' thought processes, stepping from goal to goal, does not reveal content of designing.

Tomiyama pointed to two central issues. Do we see semantical information as the driving force of design? Do we expect a computable answer? He was not interested in extracting physical properties of objects, nor processes. Culverhouse responded with the observation that knowledge engineering is restricted to rationalised knowledge, and we need to concentrate on getting the syntax right.

Mac an Airchinnigh speculated on short term memory strategies and purposeful fillers, saying that in the example of chess "strategies come in with a wham"! How can this be captured? Culverhouse responded by saying that chess masters lose as well as win, they make mistakes, as also happens with pencil and paper strategies. It was not clear how computational strategies would avoid mistakes, nor what mistakes mean in a design context. Mac an Airchinnigh concluded that we might not want intelligent systems, and should avoid systems "getting in the way" of designers.

System Architecture

5

An Experience in Adding Persistence to Intelligent CAD Environments

Jean-Paul A. Barthès[†], Patrice Anota, and Kamal El Dahshan

Université de Technologie de Compiègne
Dépt de Génie Informatique
C.N.R.S. UA 817
BP 233, 60206 Compiègne, France

Abstract: The paper deals with scaling up ICAD systems using the object paradigm for large actual applications. Several features are then necessary: *persistency, consistency, concurrency,* and *locking.* The paper focuses on persistency and concurrency. The specific nature of the implementation of persistency mechanisms depends strongly on the structure of the objects. Thus, requirements concerning the object structure are examined. Several previous solutions for persistency are described. We then present our particular approach comparing it with other works, for multi-user systems requiring concurrency control as well.
Keywords: Intelligent CAD, Object-oriented Databases, Persistency, Consistency, Concurrency, Locking

1. Introduction

Some design activities are individual in nature and do not require very large amount of data, others involve hundreds of persons and require billions of information elements over long periods of time. In between stretches a continuum of various cases. We are interested in those activities requiring several designers and a large amount of data over a long period of time. They need persistency of data and concurrent accesses of users to information. A second basic assumption in our paper is that the design process is modeled using the object paradigm. Hence the question we address is how to add persistency to object environments.

§2 introduces the problem to be solved and reviews some of the solutions that have been proposed to date. §3 describes our experience with various object environments, comparing it to some other solutions found in products or in projects in the field. A small glossary is provided at the end of the paper.

[†]Also at: SGN, 1 rue des Hérons, 78184 St-Quentin-Yvelines, France

2. Persistency for Object Environments

If we agree on the importance of the object representation for handling design problems, and if we contemplate design activities involving multiple users working over long periods of time, then we need to consider some mechanism for saving data in between design sessions as well as some mechanism for allowing several users to work simultaneously on shared data. The first requirement implies the concept of *persistency,* the second the concept of *concurrency.* In practice however such mechanisms will depend strongly on the exact format derived from the actual structure of the objects we must manipulate. Hence the question can be subdivided into two parts:

- what is a 'good' structure for representing the objects needed in design activities?
- what are the consequences on the persistency mechanism?

We first give some examples of various approaches to the problem in §2.1, then we summarize the requirements for an object representation in §2.2.

2.1. Examples of persistent environments

2.1.1. Early CAD systems. The early CAD systems (still in industrial use today) implemented persistency very easily through file storage. A user would save his environment at the end of a session and reload it at the beginning of the next one. Sharing was achieved very simply by copying data into each user environment, usually from some centralized library of models. No concurrency control whatsoever was provided.

Although easily implemented the approach presented some problems when models were updated for some reason. Furthermore only one designer could work at a given time on a given project, which would lead to either delays or conflicts when putting the pieces back together (e.g. civil engineering, piping, and electrical drawings).

The early databases were organized to manipulate large number of graphical elements and had a sophisticated *spatial indexing*. Engineering technical data were stored in separate files, or, more recently, in an associated database.

2.1.2. Relational databases. Several attempts have been made at using the relational model to represent objects. Some consist of developing CAD programs as a front end to an existing relational database, and focus on the database; others consist of coupling commercial relational data bases used as a storage medium to existing object handling environments.

As mentioned by many authors the relational model itself is not well suited for handling objects [21]. Nevertheless the existence of commercially available databases is attractive enough to justify some attempts at extending them or at interfacing them to object handling environments. An experience for multimedia applications is reported in [23]. Actual relational databases need to be modified, as was done with INGRES leading to the POSTGRES prototype [25]. Rowe then developed a mixed approach using POSTGRES to store objects and OBJFADS as a programming language [20]. The object handling is done in a LISP environment and LISP is used as a cache for each user. The intended application was the design of integrated circuits [1]. However Rowe notes that "...the major problem with implementing an object hierarchy on a relational database system is the time required to fetch an object..."

Another attempt was done with JASMINE at Fujitsu, also extending the relational data model (NF2) [14]. The object environment is written in C and includes an interpreter. Finally, using Prolog as a substrate for handling objects as was done for example in [16] seems to be quite popular.

2.1.3. Extending virtual memory. A quite different approach to ensuring persistency has been taken by researchers trying to build a persistent virtual memory. The best example is the attempt by [26]. He provides a mechanism which renders the virtual memory itself persistent. His implementation of a recoverable virtual memory is based on an efficient checkpointing scheme. However Thatte also mentions that the proposed system is not intended to eliminate the need for a file system. Furthermore if several users want to work at the same time, then virtual memory has to be shared and concurrency mechanisms added to the system, which would introduce major difficulties.

2.1.4. Object-oriented databases. Finally a last class of approaches is to provide specific object managers and to build object-oriented databases. Several attempts have been made and already include commercial products like GemStone™ using a Smaltalk object structure [19], V-BASE™ [3], STATICE™ [28], G-BASE™ in LISP environments [2]. Other projects are being developed actively. Among them, ORION [17], GARDEN [22], O2, and our project LOB [6]. Such approaches intend to solve the problems of persistency and concurrency control.

2.2. Representation requirements for CAD

As mentioned previously the complexity of the persistency mechanism is directly related to the structure of the objects to be handled. It is thus necessary to review briefly the desired features of the objects we want to manipulate.

Design is a broad activity. However if we consider it from a wide angle like in [15], it extends over the entire life of the product and includes many steps like: requirement analysis, decomposition, choice of an architecture, design specifications, manufacturing, tests, documentation, evaluation, maintenance, and improvements. Katz distinguishes three aspects of computer systems needed to support the whole process: the first one for helping to create the product, the second one to verify design decisions, the third one to manage all the needed information in order to integrate the whole process. [27] also defines the main characteristics of an integrated environment for performing Intelligent CAD IIICAD) and presents the main features that a CAD programming language should have. Previously [24] had made a survey of the CAD database area and had proposed specifications for a CAD database system. Quoting from their paper:

‘‘...A CAD database system must

- support multiple engineering application programs
- support dynamic schema modification and extension
- support the iterative nature of design
- support the embedding of semantic information in the database
- not impose constraints on the sequence of design
- automatically maintain the consistency of data

- support design versions and multiple levels of details
- support representations for complex abstraction hierarchies
- support grouping and accessing of entities by multiple relations
- support a host language interface
- support embedded and ad hoc query processing
- support a variety of management utilities
- support a multiple cpu/distributed processing environment
- support concurrent users and multiple users
- support creation and access of temporary databases
- support storage and management of programs, graphics, and representations, as well as data
- support multiple representations
- support physical data independence
- support engineering transaction processing.

..."

This is a list of difficult but useful requirements to impose on future CAD systems. It diverges from the previous attitude which emphasized graphics and tended to organize the object structure around displaying requirements. Today graphics is a secondary issue in front of the need for a precise semantic content of the representation. Thus an object should have a single internal representation and be displayed with various *surface formats* according to the viewing context. In the same fashion requirements on spatial indexing have been somewhat blurred by the availability of improved hardware. However spatial indexing is still needed in applications like 3D collision detection, or when handling large volume of data e.g. for cartography. On the other hand *data sharing* brings some genuine constraints. Access conflicts in design activities occur during long transactions when a limited number of users keep objects for a long time before releasing them to the external world. Concurrency methods must solve such a problem.

We might summarize the required features for a good object representation as follows:

- an object must be flexible (dynamic representation)
- an object type may change during the life of the object
- models (classes) may vary
- the behaviour of objects may vary
- objects can be complex (composite) and share sub-parts
- objects must be transferrable among various sites
- objects must be sharable
- objects can have multiple versions

which correspond of course to some of the requirements for a good CAD database system.

It is clear that with the exception of some of the new object-oriented database systems, most of the examples mentioned previously in §2.1 as solutions to the persistency problem cannot support such features.

3. Flexibility, Persistency, Concurrency

We describe in this section our experience in trying to develop an environment including most features listed above. The experience extends over a first prototype of Object-oriented database, VORAS, from 1978 to 1988, then to its commercial counterpart G-BASE™, and also to an on going research project, LOB. We examine the issues of flexibility of the representation, persistency, and concurrency. We also mention other approaches.

3.1. Flexibility

For coping with the problem of flexibility we designed a semantic model, PDM (Property Driven Model), later used both in VORAS and in G-BASE, and which was described in previous papers, e.g. in [10]. The implementation of the model was done in a LISP environment used as a dynamic cache for handling objects. Currently it answers all the flexibility requirements listed above, with the exception of versioning. However the flexibility achieved in VORAS and in G-BASE does not automatically follow from the choice of implementing a semantic model in a LISP environment. It involves selecting carefully the features of the model and need customizing of the environment. PDM supports many types of values as shown on a screen copy (Fig. 1) of a C3I application built by the CPM (Centre de Programmation de la Marine: French Navy) displaying text, graphics, and pictures [8].

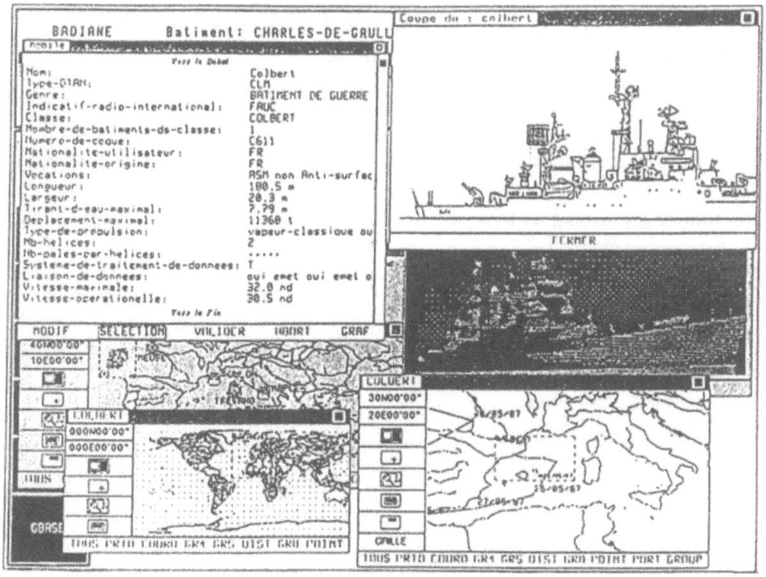

Fig. 1. Screen copy of a C3I project using G-BASE (courtesy of Graphael)

Objects in PDM are represented simply as attribute-values pairs. Some attributes have immediate associated values and are called *terminal* properties, others are links between objects and are called *structural* properties. Attributes, or properties, are in turn modeled as objects, which allows one to easily specify their behaviour. Attributes can be shared (factorized) among objects. PDM has some specific features:

- Contrary to most models, values associated with terminal properties are not restricted to any specific type, and can thus be anything from a string to a picture or an executable function (which is useful for implementing demons). However they can be restricted by the use of demons attached to the specific terminal property thus implementing data types. Such demons are invoked at creation or modification time (formatting demon) or for displaying values.

- Terminal properties may be inverted, creating external references to objects called *entry-points.* Such inversions occur at creation time, done by an appropriate demon. Entry-points are automatically maintained and updated when values are modified. Entry-points are not required to be unique, which simplifies indexing.
- Comparison operators may be user-defined and associated with terminal properties.
- structural properties are automatically inverted, which allows one to navigate easily within the database, and avoids extra indexing structures as in [18],
- Entities (all objects, classes and instances) have an internal unique identifier, their internal key. Such an identifier is otherwise called 'objid' [20], OOP [18], or UID [22].
- Any object can grow or shrink at any time.
- Classes are modeled as objects providing an integrated data dictionary. Objects are instantiated using the classes in a similar way as in Smaltalk.

PDM is a *consistent object system,* i.e. every object has a unique identifier, and there are no dangling references.

Inheritance hierarchy: As in most frame systems classes belong to an inheritance hierarchy that is orthogonal to the three level of objects: instances, classes, and metaclass (see Fig. 2). Thus PDM is more in the line of traditional AI frames like UNITS [11] or SRL Schemas [12], rather than of E-R semantic models as in [13].

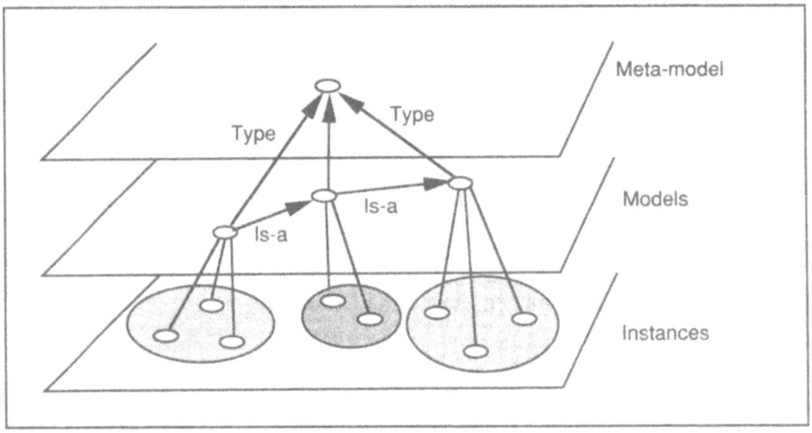

Fig. 2. PDM orthogonal specification and inheritance hierarchies

For developing an environment for ICAD, called OPAL, we used the LOB (Local Object Base) system which is a particular (new) implementation of an object-oriented database, as shown in Fig. 3.

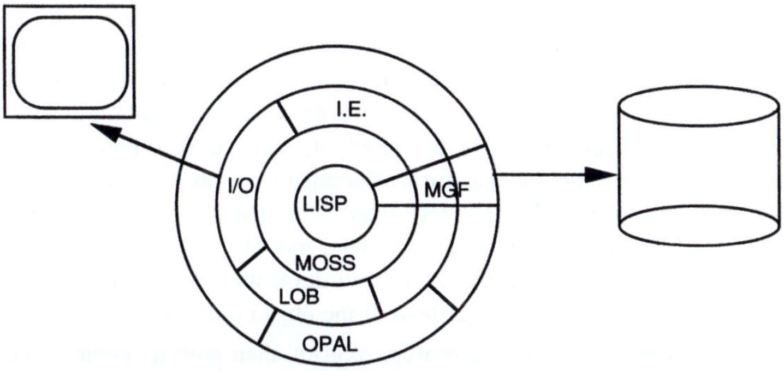

Fig. 3. Structure of the OPAL system

The LOB project uses PDM2, an extension of PDM in which

- terminal properties are *multi-valued*
- structural properties can have *multiple inverses*
- demons have been replaced with *functional objects* (methods)
- inverse properties are first class objects
- entry points are first class objects.

Furthermore LOB is implemented on top of a new object-oriented system, MOSS (Multi-level Object Simulation System), which has the following characteristics:

- messages are transmitted using *functional* calls (send)
- methods can be attached to *instances* as well as to classes or metaclasses
- there are two types of methods
    1. *own-methods* that apply directly to the object to which they are attached
    2. *instance-methods* which apply to instances of the class to which they are attached
- there are *universal methods* that are not attached to any object
- inheritance is *multiple* and flexible for universal methods
- inheritance *can be redefined* at user's will
- methods are compiled incrementally whenever they are called to improve future calls

Hence flexibility is ensured by the implementation of the PDM semantic model in a LISP environment containing an object oriented system. Furthermore, since PDM is a *reflective model,* properties themselves are first class objects and can be modified dynamically like any simple instance.

3.2. Persistency

Since we have to handle a large number of objects (eventually needing more space than the size of the virtual memory) we need to keep objects in secondary storage and to load them into working store only when necessary. However LISP environments are well known for their poor access methods (streams) obliging us to resort to back end databases for storing data. Such a solution is not well adapted when the number of objects is very large and when one needs to access single objects scattered throughout the disk space. In addition conventional relational databases display a poor behaviour when one wants to add dynamically new attributes, or to create new individual links among objects. To solve the problem, our LISP environment was extended in 1978 by adding a new access method for storing flattened structures onto disk, with direct access to each object and described in [7]. The technique consisted of transforming LISP data into ASCII forms (see fig 3.1.3) and storing the result (data, models, and functions) in a single file, while allowing multiple users (i.e. multiple LISP environments) to store and to load single objects, to and from this file simultaneously regardless of the object (unknown) *length*.

Several years of usage, first with the research prototype VORAS, then with its commercial counterpart G-BASE, showed that, although adequate, the technique involved expensive reformatting due to the many symbols describing the objects to be stored. On the other hand the particular structure of the objects allows much better performance provided some care is taken for defining a specific disk format.

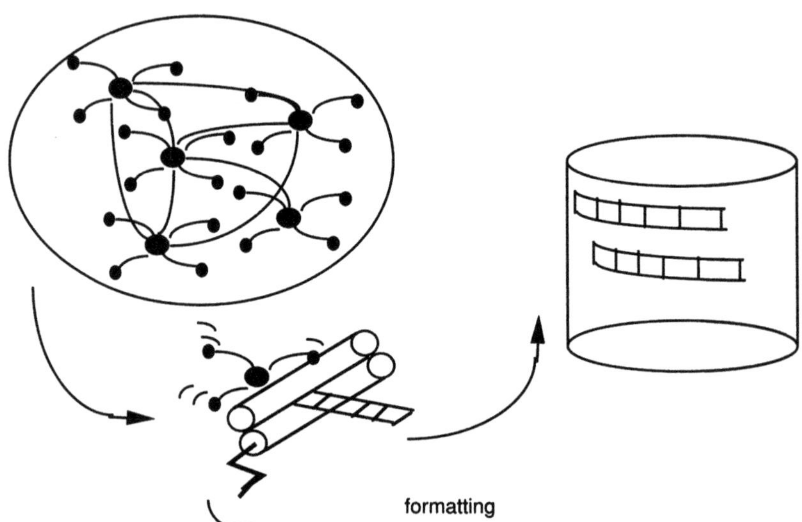

Fig. 4. Change in object format between memory and disk

We redesigned recently the object in main store and disk formats, which led to spectacular improvements. The overall results are good enough so that we can replace the need to interface LISP environments with conventional database systems (an inefficient construct), with new more flexible and integrated systems which retain the dynamic aspect of LISP while adding traditional database management features such as protection, permanence, and multiuser sharing [4].

Optimizing the format of the entity (making it as compact as possible) will in fact reduce the number of disk accesses. However we wanted to avoid a too sophisticated form for compacting data which could penalise the loading time. We had to find a compromise between astute coding and the shortest decoding time, i.e. between the gains resulting from saving bytes and the cost of restructuring the data [9].

We must also keep in mind that we are favoring the reading time rather than the writing time of the entity. Indeed, the frequency of disk reads (i.e. requests, consultation of information) by far outweighs the frequency of updating information (WORM: Write Once Read Many [25]). Therefore we can eventually sacrifice coding time in order to facilitate decoding.

Hence objects are coded in virtual memory as simple LISP a-lists (or alternated list on the LISP workstations) and must be converted to a linear format (flattened) when stored onto disk. The disk format is chosen in such a way as to minimize reading reformatting time. Property values can be stored using a binary, QFASL or XFASL standard LISP format. Furthermore the unique objid of an object can be used as the storage key. More details are available in [4].

Some other systems like STATICE or ORION [17] use a such a dual core/disk format (Fig. 4). This however is not the case for GARDEN [22] which handles objects in a flat format directly within the buffer.

In all systems, objects are stored as records on disk. They can be stored contiguously as in VORAS, G-BASE or LOB, but then disks must be packed from time to time; or they may be scattered on pages of disk like for STATICE, or GemStone, when they become too big to fit on a single page or on a segment element. In the latter case indexes must be maintained, which may be quite a complex process. The complexity is increased when the system allows for multi-users, since concurrency mechanisms have to be added.

3.3. Concurrency

Concurrency is a difficult question and a good discussion can be found in [17]. It must synchronize all accesses to a shared database preserving the consistency of data, and trying to minimize the effects of parallel accesses. Traditional mechanisms are described in [9]. However for interactive (intelligent or not) CAD systems the problem is more complex due to the existence of objects (rather than tuples in a table) and due to the unusual length of transactions (several minutes). The traditional approach for solving concurrency problems has been to use locking. We implemented a standard 2-phase lock mechanism in a version of G-BASE [5]. This is also the technique used in STATICE.

GemStone, on the other hand uses rather an optimistic concurrency control through shadowing, checking conflicts at commit time. This can be dangerous, since it could lead to aborts, and a user could loose most of the work done in a session. However work was being done to introduce some amount of pessimistic concurrency control through locks.

A problem that arises while using locking is the *level of granularity* to be used for setting the locks. Most implementations use a page lock or a segment lock, which also locks other objects contained in the same page or segment. We feel that locks are desirable at object level, which was done in the case of G-BASE, and, even so, since objects contain many links to other objects one can in practice obtain a large number of aborts. Furthermore, the locking strategy can become quite complex when composite objects hierarchies are in use like shown in [17].

Other strategies used in CAD or for CASE applications consist of creating independent versions for each user, thus eliminating de facto concurrency control. However one then has to reconcile parallel versions of a same older object, which is an unresolved problem in general.

In VORAS and in LOB, we took a different approach. Each user builds within her environment an update program, which is an easy task (of course this is done automatically). Then the program is executed serially for updating the disk. This does not implement a concurrency control mechanism. However, for some types of applications for which fancy integrity constraints (if any) can be checked at update time, it can be used efficiently. The approach is of optimistic nature, since the user can be aborted in some (rare) cases at commit time.

Other projects like GARDEN [22] do not use locks but instead use *notify triggers*. Processes communicate via mailboxes. However, since deadlocks can happen, deadlock detection is run periodically, and, if a deadlock condition is found, then some process is thrown out (dequeued).

Summarizing, there is today no magic solution to the difficult problem of concurrency control while using objects. Probably an interesting solution would be one mixing the deferred update mode of VORAS or LOB, with *posting* (broadcasting) and using context to discard irrelevant messages. We also proposed in the case of PDM to add *semantics locks,* and to use *sub transactions* while dealing with structural properties.

4. Current Status of the OPAL Project

The OPAL constraint propagation system for mechanical assembly [10] was originally written using PDM on top of an object-oriented environment BOSS. It is currently been rewritten using the PDM2 in the LOB environment, developed on top of the new MOSS object-oriented programming system. The rewriting is not completed as of today.

5. Conclusion

In a previous paper we had reported our efforts in using *object centered representation, object-oriented languages,* and *constraint propagation,* for helping the designer in the context of *mechanical assembly problems.* A major criticism of such an approach was the lack of *efficient disk storage.* Storing objects brings up a number of problems related to persistency and concurrency. We have encountered some of them while reimplementing the OPAL system on top of the LOB system (an object-oriented database) itself written using MOSS, a new object-oriented programming system in order to obtain a *seamless shared environment.*

We have listed in this paper some of the requirements for such environments handling objects for intelligent CAD systems. Some of the corresponding problems have been mentioned, and the examples of some systems implementing persistent storage have been given. Much work remains to be done in particular for improving the locking mechanisms in a demanding CAD environment. The next coming years should bring interesting new solutions for this crucial problem.

Glossary

Some simple definitions are given for terms not necessarily familiar to designers.

Persistency:

relates to considering that data remains available in between sessions thanks to some storage mechanism.

Concurrency:

implies that several designers want to access shared data. some control must be provided such that each is permitted to work as if it was alone.

Optimistic concurrency control: each user is allowed to modify hoping that there is no conflict at update time (otherwise some of the work will be lost).

Pessimistic concurrency control: if one user wants to modify some data then nobody else is allowed to obtain such data, until the first user releases it. This may lead to long waits or to 'dead locks' (A waits for B that waits for A).

Locking:

is a specific technique consisting of marking shared data in order to control access by more than one user (this is used extensively in commercial databases).

2-phase lock: a standard specific method for locking.

Consistency:

relates to the necessity of organizing a large number of objects in some coherent fashion, although the overall state of the project is not consistent (consistency requirements may be quite different in large and small problems).

This must not be confused with the consistency of the representation system (e.g. no dangling references).

Spatial indexing:

mechanisms allow the efficient retrieval of objects located in a specific volume of space. It is needed by the displaying algorithms when the number of objects in an application is large.

Transaction:

is a series of operations taking a database in a consistent state, doing some operations, and leaving it in a new consistent state.

A transaction may be aborted in which case the database is restored to the state previous to the start of the transaction. A transaction which succeeds is said to be committed. All changes are made permanent.

Reflective model:

is a representation model which can describe itself.

References

1. L.A. Rowe and C.B. Williams, *An Object-Oriented Database Design for Integrated Circuit Fabrication,* Memo UCB/ERL M87/43, University of Berkeley, 1987.
2. *G-BASE 3.0,* Graphael, 1988.
3. T. Andrews and C. Harris, ''Combining Language and Database Advances in an Object-Oriented Development Environment,'' in *OOPSLA 87, ORLANDO Proceedings*, ed. N. Meyrowitz, vol. 22, pp. 430-440, Special Issue of SIGPLAN Notices, 1987.
4. P. Anota and J.-P.A. Barthès, *A Method for Implementing Objects Efficiently in LISP Object-Oriented Databases,* EURINFO, 1988.

5. P. Anota and J.-P.A. Barthès, ''Le Controle de Concurrence dans les Bases de Données Orientées Objets,'' *Journées INRIA/AFCET sur les Bases de Données Orientées Objets*, 1988.
6. J.-P.A. Barthès, ''LOB 1.0,'' Mémo UTC/GI/DI, U de Compiègne/CNRS UA 817, 1987.
7. J.-P.A. Barthès and P. Anota, ''MGF 2.0,'' Mémo UTC/GI/DI/, University of Compiègne, 1988.
8. J.-P.A. Barthès, P. Gloess, and Y. Le Noan, ''A Command and Control System Based on a Multimedia Object-Oriented Database and a Logic Programming Language,'' *Proceedings of AISIG'89*, IEEE, Washington, 1989.
9. P.A. Bernstein, V. Hadzilacos, and N. Goodman, *Concurrency Control and Recovery in Data Base Systems,* Addison Wesley, 1987.
10. K. El Dahshan and J.-P.A. Barthès , ''Implementing Constraint Propagation in Mechanical CAD Systems,'' in *Intelligent CAD Systems II – Implementational Issues*, ed. V. Akman, P.J.W. ten Hagen, and P.J. Veerkamp, pp. 217-227, Springer Verlag, Berlin, 1989.
11. R. Fikes and T. Kehler, ''The Role of Frame-Based Representation in Reasoning,'' *Communications of the ACM*, vol. 28, no. 9, pp. 904-920, 1985.
12. M.S. Fox and J. McDermott, ''The Role of Databases in Knowledge-Based Systems,'' in *On Knowledge Base Management Systems*, ed. M.L. Brodie, and J. Mylopoulos, Springer Verlag, 1986.
13. P.C. Gilmore, ''A Foundation for the Entity-Relationship Approach – How and Why,'' in *Proceedings of the Sixth Conference on Entity-Relationship Approach*, ed. S.T. March, pp. 77-95, North Holland, 1987.
14. H. Ishikawa, F. Suzuki, and A. Makinouchi, ''Object-Oriented Multimedia Knowledge Base Management System: Design and Implementation,'' *Proceedings of the Second International Symposium on Interoperable Information Systems*, INTAP, 1988.
15. R.H. Katz, in *Information Management in Engineering Design*, Springer Verlag, 1985.
16. M.L. Kersten and F.H. Schippers, ''Towards an Object-Centered Database Language,'' *Proceedings of the 1986 Internl Workshop on Object-Oriented Database Systems*, pp. 104-112, North-Holland, 1987.
17. W. Kim, J. Banerjee, H.T. Chou, J.F. Garza, and D. Woelk, ''Composite Object Support in an Object-Oriented Database System,'' *OOPSLA 87, ORLANDO Proceedings*, vol. 22, no. 12, pp. 118-125, Special Issue of SIGPLAN Notices, 1987.
18. D. Maier and J. Stein, ''Indexing in an Object-Oriented DBMS,'' *Proceedings of the 1986 International Workshop on Object Oriented Database Systems*, pp. 171-182, Computer Society Press, IEEE, 1986.
19. GemStone Product Overview, *GemStone version 1.0,* Servio Logic Development Corporation, 1986.
20. L.A. Rowe, ''A Shared Object Hierarchy,'' *Proceedings of the 1986 International Workshop on Object-Oriented Database Systems*, pp. 160-170, Computer Society Press, IEEE, 1986.

21. J. Rumbaugh, "Relations as Semantic Constructs in an Object-Oriented Language," *OOPSLA 87, ORLANDO Proceedings*, vol. 22, no. 12, pp. 466-481, Special Issue of SIGPLAN Notices, 1987.

22. A.H. Skarra, S.B. Zdonik, and S.P. Reiss, "An Object Server for an Object-Oriented Database System," *Proceedings of the 1986 International Workshop on Object Oriented Database Systems*, pp. 196-204, Computer Society Press, IEEE, 1986.

23. K.E. Smith and S.B. Zdonik, "Intermedia: A Case Study of the Differences Between Relational and Object-Oriented Database Systems," *OOPSLA 87, ORLANDO Proceedings*, vol. 22, no. 12, pp. 452-465, Special Issue of SIGPLAN Notices, 1987.

24. S.M. Staley and D.C. Anderson, "Functional Specification for CAD Database," *Computer-Aided Design*, vol. 18, no. 3, pp. 132-138, 1986.

25. M. Stonebraker and L.A. Rowe, "The POSTGRES papers," Memo UCB/ERL M86/85, University of Berkeley, 1987.

26. S.M. Thatte, "Persistent Memory: A storage Architecture for Object-Oriented Database Systems," *Proceedings of the 1986 International Workshop on Object Oriented Database Systems*, pp. 148-159, Computer Society Press, IEEE, 1986.

27. B. Veth, "An Integrated Data Description Language for Coding Design Knowledge," in *Intelligent CAD Systems I – Theoretical and Methodological Aspects*, ed. P.J.W. ten Hagen and T. Tomiyama, pp. 295-313, Springer Verlag, 1987.

28. D. Weinreb, N. Feinberg, D. Gerson, and C. Lamb, "An Object-Oriented Database System to Support an Integrated Programming Environment," *Symbolics Inc.*, 1987.

6

Application of History Mechanism in Architectural Design

Tapio Takala and Pär Silén†

Helsinki University of Technology
Department of Computer Science
SF-02150 Espoo, Finland

Abstract: An experimental CAD system for conceptual design, called MetaViews, is described. It is intelligent in the sense that knowledge about relations (especially about causal dependencies) between design objects is stored and employed. Based on its history mechanism, containing all steps of the design process, MetaViews facilitates flexible interactive experimentation and variation of geometric forms. A case study about designing small houses with the system is described. The system's capabilities and limitations in this task are evaluated. Further development is outlined based on a fictional continuation of the same design.
Keywords: CAD, Knowledge, History Mechanism, Architectural Design, User-Interface

1. Background

With conventional CAD systems it is possible to construct a design product model, which can be analyzed and visualized by computer. As a product document a single model of the final design may be sufficient, but in the conceptual ideation phase designers often return to their previous designs, reusing and modifying earlier ideas. Then it is important to have an explicit model of the design *process* [6], containing such metalevel knowledge about the design as steps through which a design was done and reasons why each decision was made in a particular way. Based on this observation, we have developed a prototype CAD system called MetaViews [7, 8]. It has so far been applied, but is not restricted, to artistic industrial and architectural design – in particular to the early, creative sketching and ideation phases, where flexibility is of utmost importance.

A description of how an object is designed can be captured with a *history mechanism.* Perhaps the simplest way is to take sequences of commands from a log-file stored during an

† Technical Research Centre of Finland, Laboratory of Urban Planning and Building Design, Itätuulenkuja 11, SF-02100 Espoo, Finland

interactive design session. When modified and re-executed, such a sequence acts as an automatically formed parametric model for variational design [5]. In order to be really useful in designing, the history should be well-structured. Also we need a good user interface with informative representations and practical tools to manipulate the history.

There are several ways to implement a structured history. The causal dependency relations between phases of design can be understood as logic predicates implementable with logic programming [10], or as parametric transformations described with special language constructs [4], respectively. In our system, the design process is modeled as a directed network consisting of *design transactions,* each of which is a frame describing a discrete design operation. The transactions are partially ordered by links representing dependencies between them.

The MetaViews system can form the history network automatically *bottom-up* by collecting into the history the operations performed by the designer. Alternatively the user can *plan* it directly *top-down* with abstract metadesign operations. *Retrospective planning* combines both methods: a rough network is first formed by example, and is then refined with abstract planning. This way both concrete and abstract intentions of the designer can be captured by the system.

A design often consists of many *subdesigns* (e.g. parts of an assembly), each of which can be separately specified and solved. These are represented with subnetworks or structured transactions, which may be nested hierarchically.

The transaction network is a *metamodel* of the design, describing with relations the process of how design objects are formed. The system is *process-centered,* meaning that not only the design objects themselves but also processes describing their construction *(metaobjects)* are handled in an object-oriented way. Variational design can be seen as data flow programming: just put new parameters in, re-execute the process, and collect new design variations from the system's output! The same approach is also utilized in the user interface: from any design and/or analysis operations the user can compose generalized viewing processes, called *monitors.* They automatically perform the specified transformations and conversions of models, supplying additional information and reducting unessential, as is necessary in order to represent the parameters and results of a design in the most appropriate form.

In the following section, we describe how these issues are technically handled in our system. In §3 the system's flexibility in architectural design is demonstrated with examples. And lastly in §4, potential further development is outlined.

2. Features of the MetaViews System

MetaViews is an experimental CAD system mainly intended for sketching purposes. It is not comprehensive for final product documentation with sophisticated rendered pictures or fully dimensioned drawings. It rather acts as an idea processor, in which a raw model can quickly be produced and then transferred to other systems with better detailing facilities. It provides a new dimension to conceptual designing by making the design process explicitly available to the user. The system has been implemented in C, on an Apple Macintosh, which provided an easy environment for building graphical user interfaces.

The basic design tools of MetaViews are similar to those of any other CAD system. It is possible for example to interactively input and reshape polylines and polygons, to round their corners, to grow or shrink polygons with a given offset, or to perform 2-dimensional set operations (union, intersection, difference) on them. Bezier and B-spline curves can be evaluated from a control polyline as approximative polylines with a precision given by the user. 3-dimensional surfaces and solids can be produced by sweeping (translation or rotation) a 2-dimensional object in space, or by trimming (cutting by plane) another 3-dimensional object. Objects can be copied and repositioned with linear transformations (translation, rotation, scaling). Queries about objects' properties (dimensions, area, volume, etc.) will produce active text objects, which can be pasted into the design – a feature yet uncommon in CAD systems.

2.1. The history mechanism

While performing a tool operation, the system automatically stores a transaction frame describing the tool command and its numerical parameters (Fig. 1) into the history structure. Such a frame exists for each past or present design object, thus making a one-to-one correspondence between objects and design operations (they are just different facets of the same entity). Each transaction has links to all argument transactions whose results were used by the operation. Also a corresponding inverse link is added every time the transaction is used as argument by a new operation.

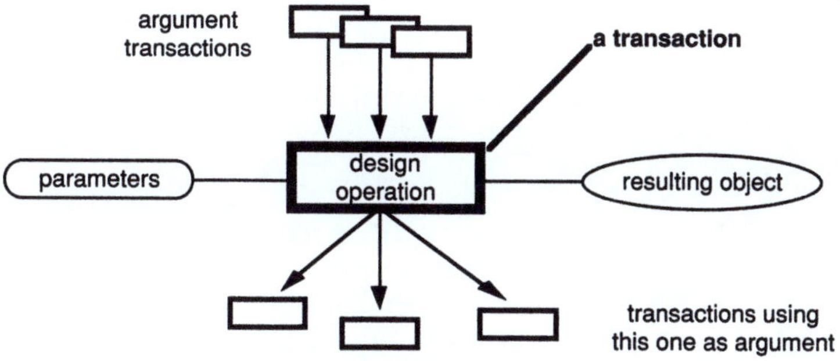

Fig. 1. A transaction frame and its associations

All transactions together form the history of a design. It is a data base with network structure formed by the links associating frames to each other. These associations are utilized as access paths to objects within a design project. Following them backwards, it is easy to find either a transaction's (e.g. no. 4 in Fig. 2) immediate arguments (no. 3) or its whole private history, i.e. the antecedent operations upon which the object is directly or indirectly dependent (nos. 1 to 4). Following forwards, we can similarly find the operation's (no. 4) immediate users (nos. 5, 11 and 11a), or all descendant objects dependent on the operation (nos. 4 to 7, 11 and 11a).

The history of an object, i.e. the sub-network formed by the corresponding transaction and all its antecedents, is a parametric metamodel. Changes made to any of its parameters are automatically propagated – through intermediate stages – to the object itself, and further to all its descendants. If a design with the same logical structure but with parametric variations is utilized several times, it can be defined as a macro operator. The object's history is extracted and a separate copy of the subnetwork is instantiated for each variation. For example in Fig. 2, history of the chair's seat part (nos. 8 to 10) is copied (forming nos. 8a to 10a) and some form parameters are changed, yielding another chair with a different seat but legs similar to the original (no. 11a).

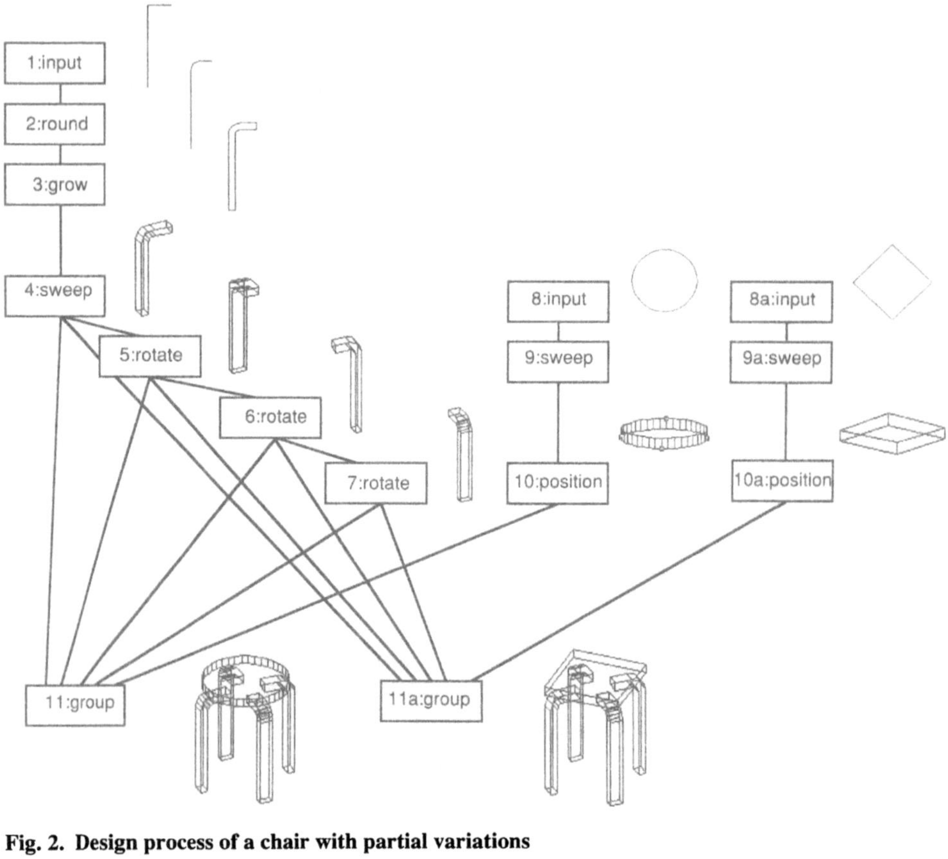

Fig. 2. Design process of a chair with partial variations

The history can be manipulated in MetaViews with *metadesign* commands. Like usual design commands, these are executed by pointing graphically to an object (in any view) and then selecting the command from a menu. However, they usually don't cause new entities to be added in the history, but rather affect the network's structure or the contents of its individual transactions. Metadesign commands for manipulating a design's history network are among others:

- *ExtractDerivation:* find from the history all antecedent operations of a selected object.
- *Collapse:* collect the selected nodes of a network into a group, which will be shown as one macro operator in the original network. Its contents can be looked at in another view.
- *CopyHistory:* create new transactions in the network, repeating the same operation with the same arguments as had those selected for copying.
- *VariateParams:* modify some parameters of a transaction, thus variating the resulting object.
- *ChangeArguments:* make an operation to act on different arguments, by redirecting the links pointing to argument objects. The network structure will thus be changed.
- *Re-evaluate:* redo a command whose argument objects or input parameters have been changed. Re-evaluation is usually done automatically for the descendants of a modified transaction, which are thus affected by the changed one.

2.2. Interactive manipulation of designs

Entities can be seen on the screen in different views. A *view* means a process consisting of a window on the screen and methods with which objects are *projected* into the window or are manipulated by input actions (here 'projection' means any conversion or transformation of an object, not necessarily a geometric projection). The number of simultaneous views is unlimited, and each may have a different projection (In Fig. 3 different views about how houses are designed are shown, see §3.2). A group of different views to the same objects (a multiview) is useful for example in monitoring interactive manipulation of 3-dimensional objects.

Selected objects from a history can be looked at geometrically in a graphical view. A metamodel (the whole project's history or an extracted part of it) can be seen through a specific *metaview* either as a network emphasizing graphically the dependency links, or in textual spreadsheet format showing more information about each transaction. At any time, the set of active entities is shown by highlighting corresponding objects and transactions in all views.

Manipulation of entities of a history network is based on the *hypermedia* concept [1]. Hypermedia facilitates metaphorical communication, allowing the user to refer to design objects via symbolic representations. An entity is selected by pointing to its representation within a view, and its associations to other entities can be followed simply by clicking a button. This kind of interactive navigation within a CAD data base, utilizing active iconic symbols with attached associative knowledge, is termed *HyperCAD* [9].

Double-clicking a transaction in metaview or a geometric object in graphical view will open it, representing either its internal structure in a metaview (if it is a collapsed macro operator) or its numerical parameters in a panel where they can be edited. The dependencies between transactions (the links to and from arguments) can be modified by moving them graphically in a network view.

Traditional bottom-up *construction* operations of CAD, performed by selecting argument objects from views and tool operations from menus, will automatically produce new transaction sequences in the history. In order not to loose information of the whole design process, every step is stored separately. When an object is changed, its previous form is not destroyed but a new

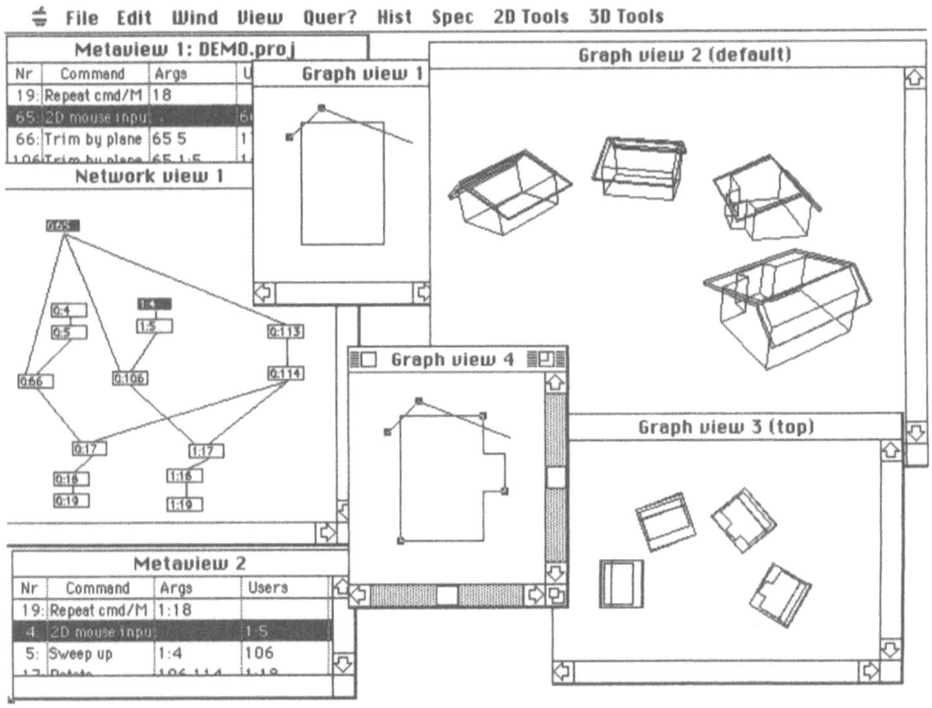

Fig. 3. A typical screen lay-out of MetaViews

modified copy is added to the history. Despite that, the user may have a feeling of directly manipulating the object, as its graphical representation on screen is immediately replaced with the new version.

Alternatively, we can *plan* a design process top-down by pasting so far unspecified operations on a network in metaview. Their exact connections to other operations and the values of their parameters can be specified later. An operation may even be an anonymous macro, which is later refined with an additional subnetwork.

The associations between entities are important hints when trying to capture the designer's intentions and way of thought [2]. For this purpose not only the design objects but also constituents of the design environment (windows etc.) are considered entities in the history. When parameters of an operation are given graphically, they become dependent on the view in which they were given. If the view's projection is changed, the parameters may become different (for example, a plane in 3-dimensional space is most naturally defined by extruding a line, drawn in the projection plane, into the viewing direction). An additional benefit of storing the whole working environment together with the design process is the possibility to suspend an uncompleted design session and later resume exactly the same situation from a history file.

3. Experiences in Architectural Design

MetaViews is experimentally applied to architectural design as part of the project ''Aesthetically Qualified Environment and New Design Technology'', where new CAD techniques are surveyed and their effects on the quality of built environment is researched.

3.1. The nature of architectural design

The behaviour of an architect doing design is often described with the terms generate and test. According to this description of the process, the architect generates a set of solutions to the problem at hand and tests their properties in terms of functional and aesthetic qualities. The testing of different proposed solutions can either be immediate subsequent to generation, or can be postponed until a set of proposed solutions have been created.

In the above mentioned research project another important process in the task of architectural design has been identified. This process could best be described with the terms focusing, subdivision and substitution.

At any moment of the design process, the designer is focusing his attention on one problem or a limited set of problems to be solved. The term focusing is important in so much as the designer, while doing this, does not omit the context of the problem(s) at hands, but keeps it in his mind as a 'fuzzy' set of criteria that affects the problem(s).

In solving the problem(s) that the designer is focusing his attention on, he usually performs two kinds of tasks. Firstly, he subdivides the problem(s) further into a set of smaller problems. Secondly, while doing this he defines the physical and functional limits of the solutions to these subproblems which, when solved and taken together, will form the solution of the actual problem he is tackling. Usually this definition of limits is done by sketching some lines on tracing paper and perhaps by making annotations in a notebook about special properties of these subproblems and proposed solutions.

The more experience the designer has, the rougher the sketches can be. An experienced designer knows how much space he must allow on the critical places to be able to later develop a functioning detail that solves the problem as a whole, while an inexperienced designer must take a bottom-up approach in every subtask. He must explore the subproblems in greater detail before he can determine whether the solution will fit into the limits of the context.

When the solving of a set of problems has reached a sufficiently detailed stage, the designer uses the solutions on all corresponding places. He substitutes the problems with the solutions. It is, in our opinion, in this process of focusing, subdivision and substitution, that the generate and test pattern of behavior is practiced over and over again.

We must emphasize, however, that this process by no means is a technical or rigid process of hierarchical subdivision. Mostly it is more like a game of associations and movements between different kinds of problems of different scales. We are convinced, that no optimal path can be scheduled, that would fit all kinds of design tasks, even if we would limit the scope of tasks to a minimum, e.g. the design of one-room log cabins. The actual order in which the designer solves the problems inherent in the design project will always be dependent firstly on the designer and his experience, secondly on the kind of artifact that is to be designed and thirdly on the context into which the artifact is to fit.

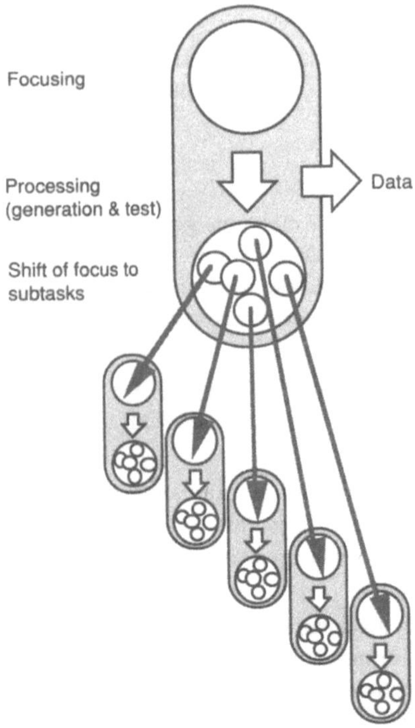

Fig. 4. The "atoms of architectural design"

It is against this underlying view of the design process that the MetaViews system is tested. The main goal is not as much to test how well the tools implemented so far can be used to design, as how the underlying philosophy of the program could serve the design process. The actual test is described below.

3.2. The objective of the experiment

The system is used in the 3-dimensional composition of building volumes. This early sketching phase of building design requires flexible modeling and easy modification of design proposals. It may start from scratch, or from a program of spaces made separately with an expert system. After having completed a sketch with MetaViews, the resulting model can be transferred into a drafting system (in this case MacDraw II or VersaCad), where the final detailed and annotated drawings can be produced.

The objective of the experiment is to design a group of small houses with a common basic form, but allowing variation in some features like dimensions of the plan, number of rooms, types of roof and windows, etc. Experiences are collected and used to evaluate the system's limits and usefulness of its tools.

Of special interest are the possibilities of using the history mechanism in the development of alternative solutions to the design problem. Assessing whether the use of the history mechanism is more or less efficient than the tools included in standard CAD packages was difficult and could not be fully performed.

3.3. Description of the design process

The design process started with the definition of the outlines of the plan of one building, by means of an input tool for 2-D polygons (Fig. 5).

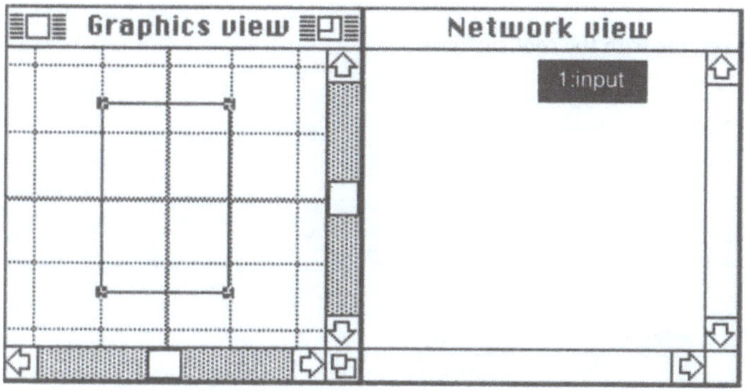

Fig. 5. The initial plan form of a house and the corresponding design history network

This polygonal form was extruded to a prism of a certain height by the *sweep* command, and the form of the roof was defined by two lines (Fig. 6).

Fig. 6. The building volume and the profile line of its roof, before the creation of roof surfaces

The two roof lines are extended into the viewing direction, resulting in planes. These are used to *trim* the prism at the roof (Fig. 7).

Thereafter thickness was added to the roof profile by the *grow* command, and the profile is swept in space for a given length, resulting in a fairly realistic wireframe view of a small building. This building was *copied, moved* and *rotated* a few times to create a group of identical houses.

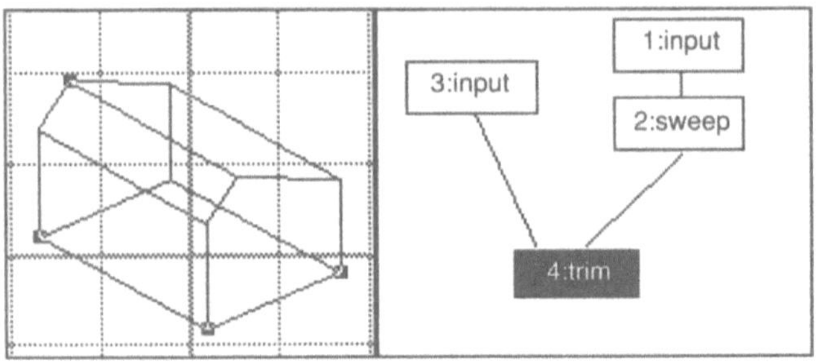

Fig. 7. Trimming the building volume with the roof surfaces

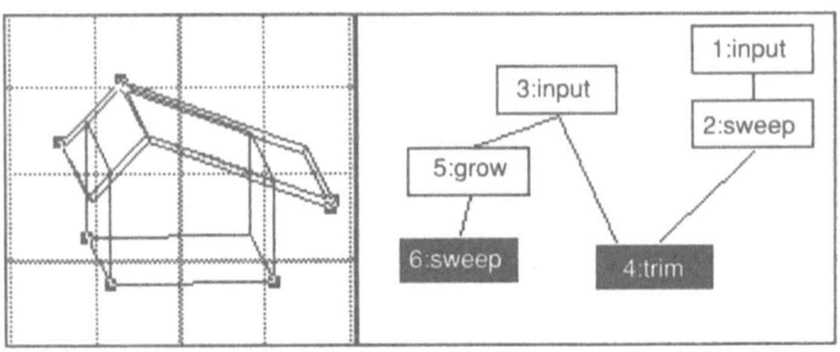

Fig. 8. A roofed house and its design history (visible parts of the network highlighted)

After this, variation of houses began with the aid of the history mechanism. The designer activated one of the buildings and *extracted* its design history in the metaview, and copied all the relevant stages of it. Among the commands in the history he activated the 2-D polygon input transaction, issued the *VariateParams* command and edited a rectangular bulge at one side of the building, depicting the adding of one room. After accepting the new form, all its descendants in the design process were *re-evaluated* using this new plan form. In a moment all the buildings, regardless of their placement, were updated on the screen, showing the same building volume (with the new room added) accurately trimmed by the old roof form.

This update did not, however, delete the old versions of the houses that existed in the history window, even though they were not shown on the graphic window. The designer activated and deleted (from view) a couple of the new designs, and activated the corresponding stages of the history of the original houses. By activating the right step in the history, and including it into the graphics view, he could instantaneously show the old versions between the new ones. In fact, all the versions were preserved and existed as alternative solutions on the same physical locations. The designer was able to choose, which of them he wanted to include in a view. The results are shown earlier in Fig. 3 (note that in the 3-D graphic views two buildings are shown from the original history, and two others from the modified branch).

In the next step of the process a new roof form was developed and applied to the two different versions of the buildings with the same success. The designer was able to freely swap between the four different kinds of buildings, retaining their physical locations.

3.4. A fictional continuation

In the following a simulated process is presented, in order to give an idea of the possibilities of MetaViews and problems that might rise due to an increased complexity in the design history network. The example makes use of two routines, *CreateWindow* and *CreatePane* that are not implemented in this version of the system. The idea is, that these routines would create windows and attach them to the wall they "belong to"

The fictional example would start exactly as the real example shown above in the Fig. 5-8. After creating the intersection of the building volume with the roof planes, the resulting surfaces are selected individually, and the situation would be as in Fig. 9.

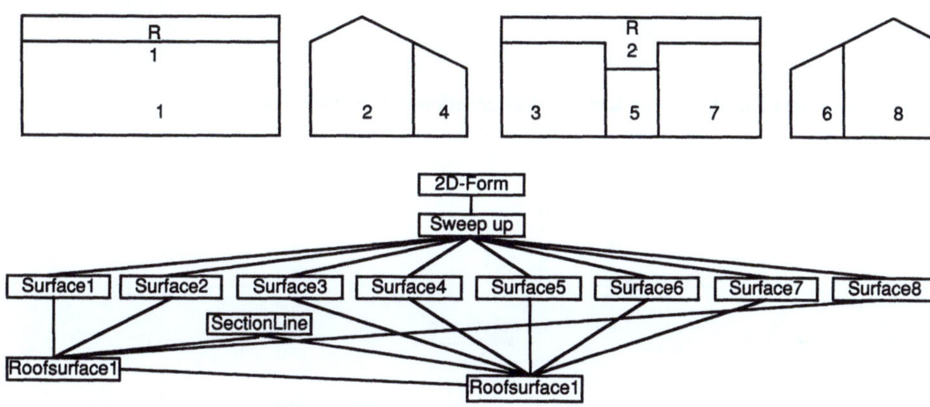

Fig. 9. The facades of a small building and the corresponding design history

Now the designer starts including windows in the design. In reality he would most probably locate the placement of the first window in the graphical view. After activating the right wall he would issue the commands CreateWindow, and twice the command CreatePane, resulting in the situation shown in Fig. 10. (The irrelevant parts of the history are dimmed for the sake of clarity.)

At this stage the normal procedure in traditional CAD systems would be to copy the window and paste it into the next place where an identical window is wanted. In MetaViews the designer can choose between doing exactly the same or duplicating the design sequence and applying it in a new context. The resulting model and design history after applying the window creating commands at wall 1 are shown in Fig. 11.

The designer could also issue another command sequence, *Copy* and *Paste* (here for the sake of clarity shown as one box). These commands create a new copy of a group of objects (the window and the panes) and position the copy in a specified place. In Fig. 12 the result of applying the Copy&Paste command to the window creating commands in the history network are shown.

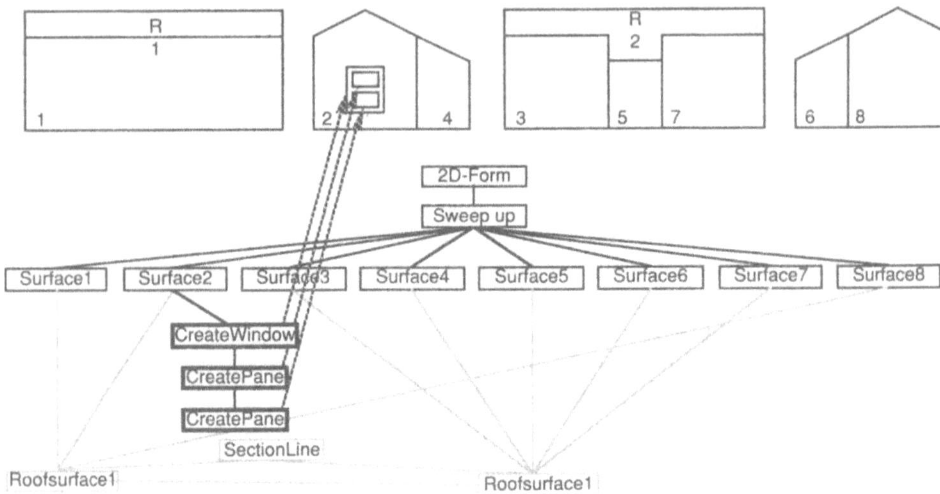

Fig. 10. The object and the design history after creation of the first window

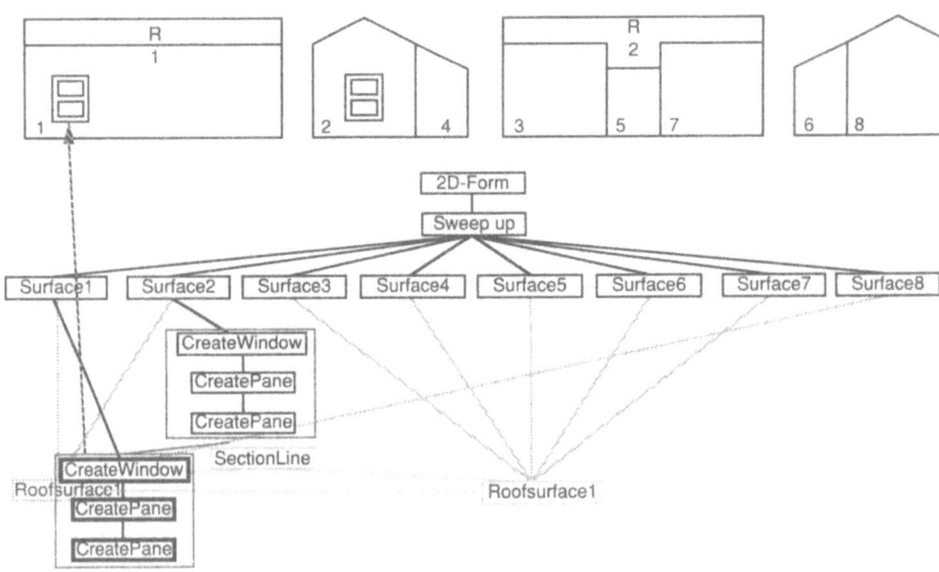

Fig. 11. The resulting model and design history

Creating two more windows on the same wall could be accomplished by applying another Copy&Paste command on both the previous Copy&Paste command and the previous window creation commands, as is shown in Fig. 13.

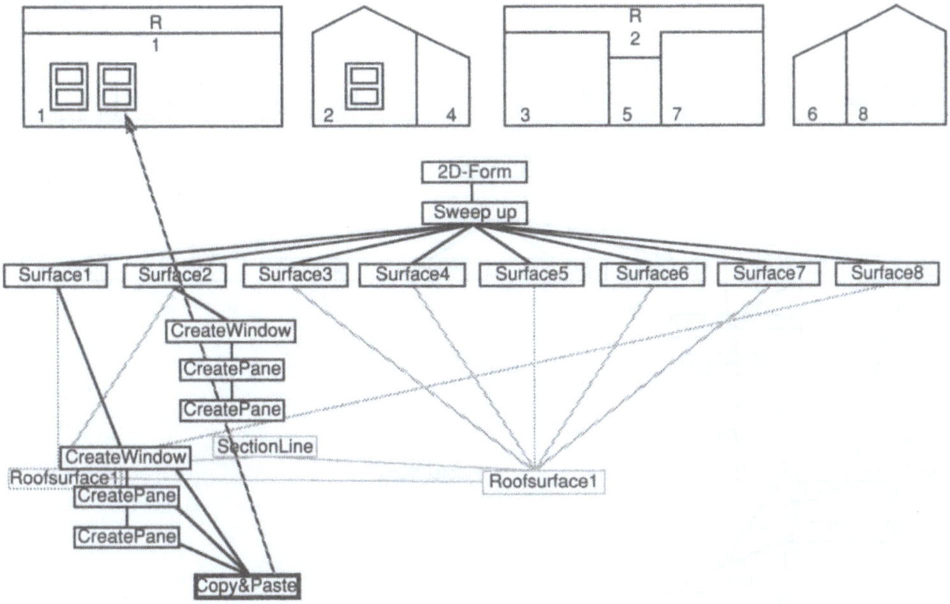

Fig. 12. The application of the Copy&Paste command

If the designer wants to create another pair of windows, but of a different size on another wall of the building, he could 'open' the CreateWindow command and change e.g. the width parameter, as shown in Fig. 14.

After this the designer could preferably apply the modified sequence of commands on the wall where he wants the new pair of windows, as shown in Fig. 15. Here, we create a pair of windows of a size different than the previously created windows by means of a modified copy of applicable nodes from the design history.

3.5. Results of the experiment

The work with the prototype so far has yielded some results, but a lot of questions are still lacking answers. A very natural obstacle, considering that the program in question is a prototype under continuing development, is the frequent interruptions caused by errors made by the user or sometimes by bugs in the code. These interruptions make it quite difficult to assess the efficiency of the program when compared to traditional modeling programs.

The most important result is in our opinion the convincing function of the history mechanism. It is quite fascinating to see how changes made to one part of the model can be pipelined to change all or just parts of the model. It could be compared to parametric design or powerful macros, with the exception that the user at no stage has to explicitly program the macros or parametric symbols.

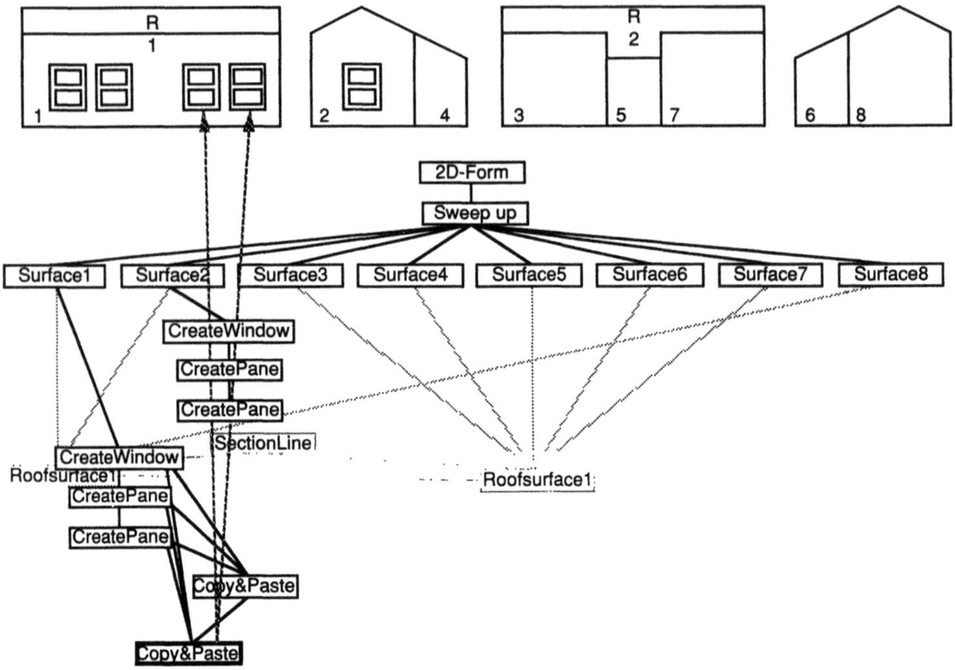

Fig. 13. One method of creating another group of two identical windows on the same wall

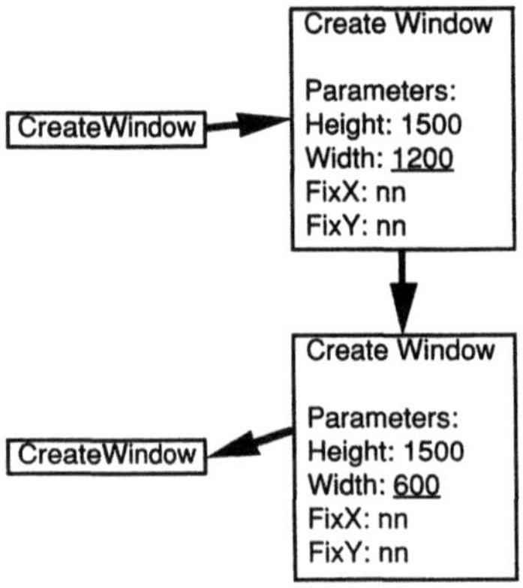

Fig. 14. 'Opening' of a node in the design history in order to change its parameters

When the user has developed different versions of his project, the strengths of the history approach become more obvious. The possibility of doing endless 'what-if' studies is very powerful in the initial stages of housing design. In practice the way of doing this is the following:

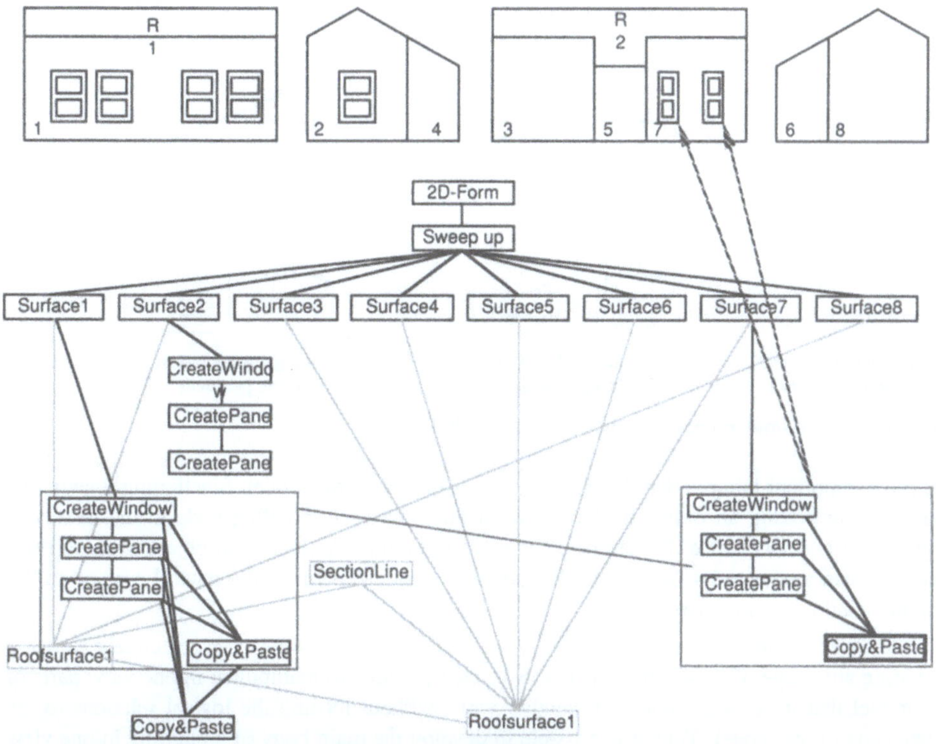

Fig. 15. The creation of a pair of windows of a different size.

The user activates the stage of the model he wants to use as the starting point of a new version. By the command *ExtractDerivation* the program activates all the stages of design that have led to this result. Then the designer duplicates the activated nodes in the history network window, and takes care that they are properly connected to other nodes. After forming the new design branch the user can change parameters, delete and insert commands or command sequences, which all affect the features of the model.

A logical problem with duplicating a history is, how the links between transactions should be directed. The schemata representing the current implementation of the CopyHistory command and a proposed solution, are shown in Fig. 16. The solution will preserve the inherent logic of the copied sequence of nodes. Currently the *CopyHistory* command copies transactions exactly, with identical links pointing to the same original arguments (see Fig. 16b). The user must change the links one by one, if the new nodes are to act as arguments. However, it might be more logical that the program automatically creates a consistent branch of such a copied history (Fig. 16c).

It is very difficult to decide which is more efficient, activating an object in the graphical view and performing actions on it, or doing the same thing with the history mechanism. Perhaps

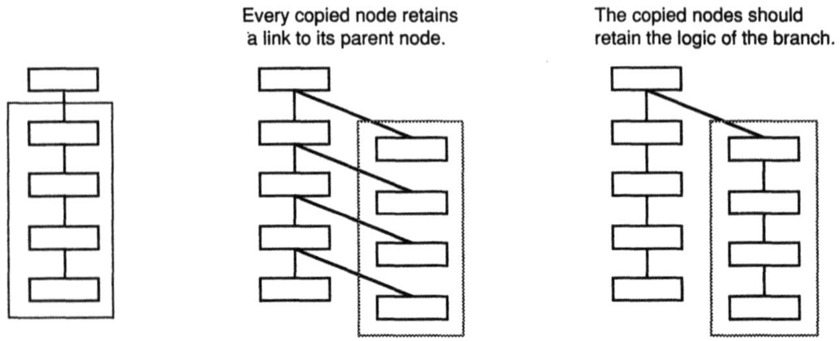

a) Original branch of a history network **b) Current function of CopyHistory** **c) Proposed solution to the problem**

Fig. 16. Current implementation and a proposed solution

the most important thing is that there are two ways to look at the model, functioning together so that when activating an object in the graphical view the corresponding node in the history is highlighted and vice versa. Sometimes it is more convenient to locate an object in the history network than among a lot of lines in the graphical view, sometimes the natural method in the graphic window is more appropriate.

Another important aspect of the history mechanism is the possibility of selectively showing or hiding any stage of the design process. It is even possible to manipulate in one view parts of the model that have been created in another view, without loosing the logical relations to the other parts of the model. Thus it is possible to develop the main body of a building in one view and the roof in another, and to see the resulting combination of them in a third view, or perhaps in multiple views from different directions.

The program, as it is at the moment of writing this, has a set of 2- and 3-dimensional tools that are barely enough to do fundamental modeling of buildings. It is naturally necessary to develop many more tools before the program could replace conventional 3-dimensional modelers. Some of the methods applied in the Schema program [3] could be very good to implement also in MetaViews in order to enhance the user-friendliness of the program. It is however clear, that the history mechanism is a very powerful fundamental philosophy, that greatly widens the possibilities of viewing and manipulating models in architectural design.

4. Conclusions

The experiment conducted in this study has demonstrated the potential of the history mechanism of MetaViews in architectural design applications. Designers find it very convenient to build parametric models without explicitly programming them. Also the ability to handle different phases of design in simultaneous views gives better insight to the design parameters and their effects. Quick variational design experiments can easily be produced with these tools.

However, as the system is still an experimental prototype, a lot of questions considering its practical efficiency are still lacking answers. With our small scale example it is not possible to investigate, how one should navigate in a large network database without getting "lost in hyperspace".

Some directions for further development can readily be pointed out. A more comprehensive set of geometric modeling tools is evidently needed. Geometric sketching also requires tools beyond traditional solid/surface modeling. For example, techniques for handling infinite or partially trimmed objects are essential. Also other relations than direct dependencies between entities, like constraints on design parameters and resulting objects, or associations to painted freehand sketches and verbal annotations would be very useful.

Acknowledgements

The original work leading to MetaViews was done at Helsinki University of Technology and has been supported by the Technology Development Centre. Practical application experiments and later development, continuing at Technical Research Centre of Finland with the project 'ESSU' (Aesthetically Qualified Environment and New Design Technology), are funded by the Emil Aaltonen Foundation.

References

1. J. Conklin, "Hypertext: An Introduction and Survey," *IEEE Computer*, pp. 17-41, 1987.

2. R. Makkuni, "A Gestural Representation of the Process of Composing Chinese Temples," *IEEE Computer Graphics and Applications*, vol. 7, no. 12, pp. 45-61, 1987.

3. M. van Norman, "A Digital Modelshop: The Role of Metaphor in a CAAD User Interface," *Design Computing*, vol. 1, pp. 95-122, 1986.

4. J. Rossignac, P. Borrel, and L. Nackman, "Interactive Design with Sequences of Parametrized Transformations," in *Intelligent CAD Systems II – Implementaional Issues*, ed. V. Akman, P.J.W. ten Hagen, and P.J. Veerkamp, pp. 93-125, Springer-Verlag, Berlin, 1989.

5. T. Takala, "User Interface Mangement System with Geometric Modeling Capability – A CAD System's Framework," *IEEE Computer Graphics and Applications*, vol. 5, no. 4, pp. 42-50, 1985.

6. T. Takala, "Theoretical Framework for Innovative Computer Aided Design," in *Design Theory for CAD, Proc. of IFIP WG 5.2 Working Conference on Design Theory for CAD*, ed. H. Yoshikawa, and E. Warman, pp. 323-334, North-Holland, Tokyo, 1986.

7. T. Takala, "METAVIEWS - A Process-Oriented Approach to CAD and Graphics," in *Proc. of Eurographics'88*, pp. 171-182, North-Holland, Nice, 1988.

8. T. Takala, "Design Transactions and Retrospective Planning: Tools for Conceptual Design," in *Intelligent CAD Systems II – Implementaional Issues*, ed. V. Akman, P.J.W. ten Hagen, and P.J. Veerkamp, pp. 262-272, Springer-Verlag, Berlin, 1989.

9. D. Vanier, "Hypertext - a Computer Tool to Assist Building Design," in *Conceptual Modeling of Buildings, CIB W74 + W78 Seminar*, Lund Institute of Technology, Lund, 1988.

10. Y. Yamaguchi, F. Kimura, and P.J.W. ten Hagen, ''Interaction Management in CAD Systems with History Mechanism,'' *Proc. of Eurographics'87*, pp. 543-554, North-Holland, Amsterdam, 1987.

7

Discussion Session: System Architecture

Ken MacCallum

1. Papers

This session contained three papers which, although being concerned with aspects of the architecture for Intelligent CAD Systems, were quite diverse in their nature. The first by Jean-Paul Barthès emphasised the problems of persistence associated with large knowledge bases. The second by Alan Shenton presented ideas on the use of function to match input and outputs of processes and the application of these ideas in conceptual design and requirements analysis. In the third paper, Tapio Takala presented an argument and approach for involving and using a design history mechanism.

When looking for a common thread between these papers and why they come together in this Workshop, an observation which can be made is that we are now examining problems which are closer to those relevant to real delivery systems. This can be no more than an observation, but does perhaps suggest the beginning of a realisation of the work of past years.

There is, however, a stronger theme which is apparent in the three presentations. In each case a central concern is to hold onto the past. For Barthès, a major issue is to make information reusable. Thus, it should be possible to allow an accumulation of design knowledge over a number of design cycles so that realistic quantities of knowledge are available. In Shenton's case, it is the applications which are to be reusable. The idea is that the processes with their inputs and outputs can act as design prototypes and be used in different design situations. The capturing of histories proposed by Takala stems from a desire to have reusable design processes, or, more strictly, design scenarios. In all three cases the phase of activity which is relevant is the full design cycle rather than just a design task. In other words, the concern is not just to produce a CAD system which is intelligent, but to ensure that knowledge used in such processes can accumulate and develop over a sequence of designs. This does in fact reflect, more closely, people's expectations of an intelligent system.

2. General Discussion

An issue which is relevant to all intelligent CAD systems, but which is highlighted by this concern for building on the past is that of consistency. It was on this issue that the major debate was carried out in the discussion session. Barthès opened up with a request for a new definition of consistency. This he argued is necessary since large databases are always inconsistent, and logic is unusable because of its strict definition of consistency. He went on to say that a knowledge base should act like a data base in which knowledge is stored independently from the inference methods. When there are several sources which supply knowledge in independent formats, this may entail inconsistency. Thus, he argued, the knowledge base needs to be partitioned so that the inconsistencies are not important or apparent. Takala pointed out that partitions of knowledge can be seen as viewpoints which are consistent; but still different views can be inconsistent.

The discussion broadened out when it was pointed out by Shenton that inconsistency is the norm in design. It exists in the mind of the designer, it is essential to our design problem solving, it can even be communicated to others without apparent difficulty. Takala prefers to think of 'converging sequences'; Eshuis prefers the term 'over-specified'. Benyon argued that you cannot exhaustively describe all of the functions of an artifact; there has to be a situation where a system is incomplete - this in itself is an inconsistency.

As the discussion developed, it became clear that there were various different uses of the word consistency. While logic has its own strict form of consistency it only goes so far as the resolution of the statements composed from strings of symbols. There is no reference as to how these symbols should be interpreted in the real world. This difference between the logical semantics and the real world semantics is quite crucial; indeed it is over-simplistic to consider that there are only two levels of semantics. Pineda makes this point in referring to the ''semantics of designing''.

Tomiyama makes a plea for some kind of mechanism for ''keeping the peace''. In other words we should abandon our attempts to define consistency, and instead have systems which can support inconsistency happily. This is an attractive concept, and although it was not feasible to take it any further in the discussion, it perhaps best represented the feelings of the meeting.

In summary, we can say that issues of consistency and inconsistency are important. However, we have to be clear as to what we are referring when we use the terms. It is quite feasible, and perhaps desirable to use a symbol system which is consistent, but whose interpretation by other processes leaves inconsistencies. If we can properly define the nature of these inconsistencies so that they are in fact consistent according to a design definition rather than a real world definition then we can start to define a ''peace maintenance system.''

Languages

8

A Graphical and Logical Language for a Simple Design Domain

Luis A. Pineda and Ewan H. Klein

EdCAAD and Centre for Cognitive Science
University of Edinburgh
20 Chambers Street, Edinburgh EH1 1JZ, UK

Abstract: In this paper a logical and graphical language for representing CAD knowledge in a simple design domain is presented. The language is useful for designing 2-dimensional wire-frame diagrams of the sort that are common in architectural and other kinds of drawings. The language allows the definition and interpretation of graphical symbols that are explicitly drawn on the screen, as well as the representation of other context-dependent space partitions that receive an interpretation in terms of the graphical context from which they emerge. For the identification of these emerging graphical objects we introduce the notion of *intension* of a graphical symbol. In addition, the language supports the definition and representation of construction lines. This facility is particularly useful for modelling causal relations between possible design states in the definition and production of design inferences.
Keywords: Semantics of Graphics, Knowledge Representation, Intelligent CAD.

1. Introduction

In this paper we show how a language for geometric reasoning for CAD with a sound theoretical foundation can be implemented in a graphics interactive environment. The resulting system allows the representation and interpretation of complex object and functions, as well as possessing a clear semantic interpretation. The language is specified as an algebraic system. The algebraic specification of geometrical data types for CAD systems has been explored with promising results; for instance, in the so-called definitive programming framework [1]. We also show the implementation of the graphical and logical language in a programming environment called GRAFLOG [6-8]. The language is formally presented in §2. In §3 we discuss the notions of extension and intension of graphical representations, and we define a criterion by which drawings are identified as representing the same object in throughout a process of change.

In §4, an interaction with the system in a simple design domain is illustrated. Here, some details of the implementation of GRAFLOG in Prolog are shown. The way basic and emergent graphical symbols are represented and referred to by the language is highlighted in this section too. In §5, a procedure for inferring the interpretation of drawings in our simple design domain –the graphical parsing procedure– is presented. For this purpose, a set of identification rules acting upon the graphical and linguistic input is defined. Finally, the definition of some design constraints and modelling rules in the course of graphics and linguistic dialogue is illustrated in §6.

2. The Graphical Structure D and the Language $\mathbf{L_d}$ for Design

For the formalisation of the structure of graphics and its integration in the logical representation, the following strategy is adopted. First, an algebraic structure **D** for graphics in 2-dimensional wire-frame domain is developed. This structure has a declarative and an algorithmic interpretation. This duality is at the heart of the theory, and the two interpretations of **D** are illustrated. Second, the structure **D** is embedded within the modal first order logical language $\mathbf{L_d}$ in which both linguistic and graphical expressions can be represented. Some examples of expressions that refer to graphical and abstract individuals, properties and relations are illustrated. The algebraic structure allows us to represent the meaning of well-formed graphical representations of discrete states of the graphics interactive session; however, the algebraic system must be extended for capturing the meaning of drawings when they undergo a process of change. For this purpose, modality is introduced in the representational system.

In the formulation of **D** we follow Goguen et al.'s approach to abstract data types [4]. We start with some general definitions. Let S be a set of sorts. An S-sorted signature Σ is a family $\Sigma_{w,s}$ of sets, for each $s \in S$ and $w \in S^*$ (where S^* is the set of all finite strings over S, including the empty string e). An operation symbol $F \in \Sigma_{w,s}$ is said to have **rank** w, s, **arity** w and **sort** s. If c is a symbol of rank e, s (i.e. where e is the empty string), then c is called a **constant** of sort s. For example, we might take our set $S = \{integer\}$, and $\Sigma_{w,s}$ to be empty except for $\Sigma_{e,\,integer} = \{0\}$ and $\Sigma_{integer,\,integer} = \{succ\}$. That is, *succ* is to be interpreted as a unary operation (i.e. successor) which takes an argument n of sort *integer*, and yields a value $succ(n)$ which is also of sort *integer*. This system produces the set of natural numbers, that is, the carrier S.

2.1. The graphical structure D

After these preliminaries, we turn to the graphical structure **D** for drawings made out of dots, lines and polygons.

(1) The set $S_{\mathbf{D}}$ of sorts = $\{bool, real, real\_pair, dot, line, polygon\}$.

The individual constants are:

(2) $\Sigma_{e,\,bool}$ is the set $\{0, 1\}$.

(3) $\Sigma_{e,\,real}$ is a (non-countably) infinite set of numerals.

(4) $\Sigma_{e,\,dot}$ is a (non-countably) infinite set $\{d_0, d_1, \ldots\}$ of dots.

(5) $\Sigma_{e,\,line}$ is a (non-countably) infinite set $\{l_0, l_1, \ldots\}$ of lines (construed as vectors).

(6) $\Sigma_{e,\,polygon}$ is a (non-countably) infinite set $\{p_0, p_1, \ldots\}$ of polygons.

The operator symbols are as follows:

(7) $\Sigma_{bool\ dot,\ real\_pair} = \{position\_of\}$.

(8) $\Sigma_{bool\ line,\ real} = \{length\_of\}$.

(9) $\Sigma_{bool\ polygon,\ real} = \{area\_of\}$.

(10) $\Sigma_{bool\ line\ line,\ real} = \{angle\_between\}$.

(11) $\Sigma_{bool\ line,\ dot} = \{end\_of,\ origin\_of\}$.

(12) $\Sigma_{bool\ line\ line,\ dot} = \{int\_oo,\ int\_om,\ int\_oe,\ int\_mo,\ int\_mm,\ int\_me,\ int\_eo,\ int\_em,\ int\_ee,$ $int\_ww,\ cross\_at,\ t\_joins\_at,\ e\_join\_at,\ joint\_at,\ intersect\_at\}$.

(13) $\Sigma_{bool\ polygon\ polygon,\ polygon} = \{union\_of,\ intersection\_of,\ difference\_between\}$.

Note that according to clause (11), there are operation symbols that associate with every line *l* values of sort *dot* which indicate the origin and end point of *l*. That is, *l* is oriented, and is in fact encoded as an ordered pair of dots. Other terms of sort *dot* are produced from the intersection of two terms of sort *line*.

Clause (12) defines a set of operator intersections. This class covers the vector relations traditionally considered in computer graphics, and also a set of intersection modes used for the definition of construction lines. For instance, a term of the form *int_mm*(*true*, *A*, *B*) denotes the dot in which the vectors *A* and *B* intersect each other as shown in Fig. 1. The first argument with the value *true* indicates that such a dot actually exist in the state of affairs that is denoted by the algebraic expression. If the vectors *A* and *B* were parallel, the intersection dot would not exist, the first argument of *int_mm*'s arity would be false, and the whole expression would have no referent in that state. We discuss further the treatment of this kind of irregularities in §2.3.

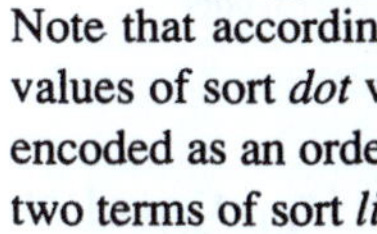

Fig. 1. Two vectors intersect each other

There are also some operation symbols taking boolean values for testing geometrical conditions, i.e. geometrical predicates:

(14) $\Sigma_{bool\ line,\ bool} = \{horizontal,\ vertical\}$.

(15) $\Sigma_{bool\ line\ line,\ bool} = \{perpendicular,\ parallel,\ collineal\}$.

(16) $\Sigma_{bool\ dot\ polygon,\ bool} \cup \Sigma_{bool\ polygon\ polygon,\ bool} = \{in\}$.

(17) $\Sigma_{bool\ dot\ line,\ bool} = \{on\}$.

The syntactic rules for the definition of well-formed expressions of **D** are given below, when the logical extension of this structure is presented.

2.2. The language $\mathbf{L_d}$

Now, we embed the structure **D** within a modal first order logical language $\mathbf{L_d}$. This will allow us to represent graphical and logical knowledge in an integrated fashion. We define the set of sorts $S_\mathbf{d}$ of $\mathbf{L_d}$ as follows: $S_\mathbf{d} = S_\mathbf{D} \cup \{individuals\}$. These additional individuals are referred to by natural language expressions, and might be physical or abstract entities. The modal first order language $\mathbf{L_d}$ is defined as follows:

(18) All operator symbols of **D** are also in $\mathbf{L_d}$.

(19) $\Sigma_{bool\ bool,\ bool} = \{\wedge, \vee, \rightarrow, \equiv\}$

(20) $\Sigma_{bool,\ bool} = \{\neg, \Box\}$

(21) *Quantifiers*: For every sort $s \in S_\mathbf{d}$, $\forall_s$ (for all), $\exists_s$ (there exists), each of rank $\Sigma_{s,\ bool}$

(22) *Functional Abstractor*: For every sort s in $S_\mathbf{d}$, $\Sigma_{s,\ bool} = \{\lambda\}$.

(23) *Equality*: For every sort s in $S_\mathbf{d}$, $\Sigma_{s\ s,\ bool} = \{=\}$

(24) *Variables*: For every sort $s \in S_\mathbf{d}$, there is a countably infinite set $\mathbf{V}_s = \{x_{s,0}, x_{s,1}, \ldots\}$ of variables such that $x_{s,i} \in \Sigma_{e,\ s}$.

(25) *Constants*: For every sort $s \in S_\mathbf{d}$, there is a countable set $C_s = \{c_{s,0}, c_{s,1}, \ldots\}$ of constant symbols such that $c_{s,i} \in \Sigma_{e,\ s}$.

(26) *Predicates*: For every sort $s \in S_\mathbf{d}$, there is a countable set $P_s = \{p_{s,0}, p_{s,1}, \ldots\}$ of predicate symbols such that $p_{s,i} \in \Sigma_{u_0\ u_1 \ldots\ u_i,\ bool}$. The string $u_0\ u_1 \ldots\ u_i$ is called the **arity** of $p_{s,i}$.

(27) *Auxiliary symbols*: '(' and ')'.

Formation rules are expressed as follows:

(28) Every constant of sort s is a term of sort s.

(29) If $t_1, \ldots, t_n$ are terms of sorts $s_1, \ldots, s_n$, respectively, and f is an operation symbol of rank w, s, where $w = s_1, \ldots, s_n$ then $f(t_1, \ldots, t_n)$ is a term of sort s.

(30) If $t_1, \ldots, t_n$ are terms of sorts $s_1, \ldots, s_n$, respectively, and f is a predicate symbol of rank w, *bool*, where $w = s_1, \ldots, s_n$ then $f(t_1, \ldots, t_n)$ is a term of sort*bool*.

(31) If $\forall_s$ is a quantifier of sort s, u is variable of sort s and ϕ is a term of sort *bool* then $\forall_s u\phi$ is a term of sort *bool*.

(32) If $\exists_s$ is a quantifier of sort s, u is variable of sort s and ϕ is a term of sort *bool* then $\exists_s u\phi$ is a term of sort *bool*.

(33) If λ_s is a functional abstractor of sort s, u is variable of sort s and ϕ is a term of sort *bool* then $\lambda_s u\phi$ is a term of sort *bool*.

2.3. Error elements and the closure of $\mathbf{L_d}$

Operations in standard multi-sorted algebraic systems denote total functions. That means that if the arguments of an operation term are of the appropriate sorts, the value of such an operation is an element of some appropriate sort as well. Consider the case in Fig. 1 in which vectors A and B intersect each other. The expression *int_mm(true,A, B)* denotes the intersection dot. Notice that the first argument in the operation's rank is a term of sort *bool*. This term is defined to be *true* or *1* if the result of the operation is of a proper sort, as it is the case in Fig. 1. However, suppose that the arguments of sort *line* of *int_mm* are two vectors that happen to be parallel. Then, the term *int_mm(false, A, B)* denotes no dot at all in relation to such a pair of vectors. In general, an operation of **D** that takes as arguments graphical symbols that have appropriate spatial properties, like their position and dimensions, has as a value a graphical object of the operation's sort. However, if the properties of the symbols that are the arguments of an operation have inappropriate values, the operation would not produce an object of its corresponding sort, despite the fact that the arguments themselves are of the right sort.

For handling irregular conditions in drawings we define an infinite number of *error* or ill-formed elements of every sort in every carrier in $S_{\mathbf{D}}$. That is, besides the 'normal' elements that belong to the carrier of a sort *s*, there are an infinite number of objects, namely $e_{<s,0>}$, $e_{<s,1>}, \ldots, e_{<s,n>}$, which act as the value of operations that otherwise would be undefined. For instance, if *A* and *B* are parallel then $int\_mm(false, A, B) = e_{<dot,i>}$. With this provision for error handling, every expression of **D** and in which every argument is a symbol of an appropriate sort denotes a total function. Errors are propagated in complex expressions; for instance, $int\_mm(false, e_{line,i}, B) = e_{<dot,j>}$.

2.4. Algebraic interpretation of $\mathbf{L_d}$

A *model M for* $\mathbf{L_d}$ is an ordered tuple $<D, I, F>$, where $D = <D_s>_{s\in S}$ is an *S*-indexed family of non-empty sets, *I* is a set $\{i_1, i_2, \ldots\}$ of states and *F* is an interpretation function whose domain is the set of all non-logical and non-graphical[1] constants of $\mathbf{L_d}$ and whose range is described below. If α is is constant of sort *s*, then $F(\alpha)(i) \in D_s$.

We define as well an assignment function *g* that has as its domain the set of all variables and has as a value a member of G_s for each variable of sort *s*. Now we define the semantic interpretation of expressions of $\mathbf{L_d}$:

(34) If α is a constant of sort *s*, then $[[\alpha]]^{\mathbf{M, i, g}} = [F(\alpha)](i)$. The extension of a constant symbol at every index $i \in I$ is the same object in D_s.

(35) If α is a variable then $[[\alpha]]^{\mathbf{M, i, g}} = g(\alpha)$.

(36) If α is a term of sort *bool* and *u* is a variable of sort *s*, then $[[\lambda u \alpha]]^{\mathbf{M, i, g}}$ is a function *h* with domain in D_s, such that for any object *x* in that domain $h(x) = [[\alpha]]^{\mathbf{M, i, g'}}$ where g' is that value assignment exactly like *g* with the possible difference that $g'(u)$ is the object *x*.

(37) If *f* is an operation symbol in **D** of rank *w, s*, for $w = s_1,\ldots,s_n$ then $[[f]]^{\mathbf{M, i, g}}$ is a function with domain in $D_{s_1} \times \ldots \times D_{s_n}$ and range in D_s. This function, for every operator symbol of **D**, is defined in terms of geometrical analysis, and is computed by its associated algorithm.

(38) If *p* is a predicate symbol of rank *w, s*, for $w = s_1,\ldots,s_n$ then $[[p]]^{\mathbf{M, i, g}} = [F(p)](i)$ such that $[F(p)](i) \subseteq D_{s_1} \times \ldots \times D_{s_n}$

(39) If $t_1,\ldots,t_n$ are terms of sorts $s_1,\ldots,s_n$ respectively, and *f* is an operation symbol of rank *w, s*, where $w = s_1,\ldots,s_n$ then $[[f]]^{\mathbf{M, i, g}}([[t_1]]^{\mathbf{M, i, g}},\ldots, [[t_n]]^{\mathbf{M, i, g}})$. The result of applying the function $[[f]]^{\mathbf{M, i, g}}$ to its arguments $[[t_1]]^{\mathbf{M, i, g}},\ldots, [[t_n]]^{\mathbf{M, i, g}}$.

[1] The graphical constants are the operation symbols of **D**.

(40) If α and β are terms of sort s then $[[\alpha = \beta]]^{\mathbf{M, i, g}}$ is *1* if and only if $[[\alpha]]^{\mathbf{M, i, g}}$ is the same as $[[\beta]]^{\mathbf{M, i, g}}$.

(41) If ϕ and ψ are terms of sort *bool* then $[[\neg\phi]]^{\mathbf{M, i, g}}$ is *1* if and only if $[[\phi]]^{\mathbf{M, i, g}}$ is *0*, and $[[\neg\phi]]^{\mathbf{M, i, g}}$ is *0* otherwise.

(42) If ϕ is a term of sort *bool* then $[[\phi \wedge \psi]]^{\mathbf{M, i, g}}$ is *1* if and only if $[[\phi]]^{\mathbf{M, i, g}}$ is *1* and $[[\psi]]^{\mathbf{M, i, g}}$ is *1*. (The definition for other truth-functional connectives is given in the standard way along the lines shown for the connective $\wedge$).

(43) If ϕ is a term of sort *bool* and u is a variable of sort s then $[[\forall_s u\phi]]^{\mathbf{M, i, g}}$ is *1* if and only if $[[\phi]]^{\mathbf{M, i, g'}}$ is *1* for all g' exactly like g except possibly for the value assigned to u.

(44) If ϕ is a term of sort *bool* and u is a variable of sort s then $[[\exists_s u\phi]]^{\mathbf{M, i, g}}$ is *1* if and only if $[[\phi]]^{\mathbf{M, i, g'}}$ is *1* for some value assignment g' exactly like g except possibly for the value assigned to u.

(45) The interpretation of the modal operator $\Box$ is such that if ϕ is an expression of sort *bool*, then $[[\Box\phi]]^{\mathbf{M, i, g}}$ is *1* if and only if $[[\phi]]^{\mathbf{M, i, g}}$ is *1* for all i in I.

Definition of truth:

(46) If ϕ is a term of sort *bool*, then ϕ is *true with respect to M and to I* if and only if $[[\phi]]^{\mathbf{M, i, g}}$ is *1* for all value assignments g.

At this point, it is worth highlighting one important difference between the way of specifying logical truth functions and functions that are named by the operation symbols of **D**. The semantics of truth functors, like $\wedge$ and $\rightarrow$, is usually defined in terms of a truth table, or as shown in clauses (42) and (43). Given that there are just three distinctive elements in the carrier of sort *bool*, namely *1*, *0* and e_{bool}, we can give these definitions in a discrete fashion. However, the domains of the functions that are named by operation symbols of **D** have an infinite number of distinctive elements. For these definitions, we rely in a background theory that is provided by geometrical analysis. For instance, the function named by operation symbols of rank *bool line line, dot* in clause (12) is defined below by the system of vectorial equations in (47). Furthermore, for each of these operation symbols there is an algorithm that for every collection of arguments of the proper sort, computes the function's value. If the arguments are not proper, an error message is produced. Such a message 'implements' an error element of the operation's sort. Each of these algorithms with its associated error messages can be thought of as computing a total function. For every operation symbol in **D**, the definition and application of such a function are respectively specified in clauses (37) and (39).

Functional application in (36) is defined in the standard way: if x is a variable of sort s, $\phi(x)$ is an expression of sort *bool* and a is a constant of sort s, the application of $\lambda x\phi(x)$ to a, $\lambda x\phi(x)(a)$, is $\phi(a)$. The introduction of λ-terms is useful for the definition of design concepts [8].

2.5. Algorithmic interpretation of D

For every operation symbol of **D** there is a geometrical algorithm that is named by that symbol. In general, the geometrical information represented in **D** is computed by standard geometrical analysis. The sorts of **D** have been selected precisely because they are relevant for computing the geometrical algorithms involved in the operations. It can also be observed that the geometric predicates, i.e. the operation symbols of sort *bool*, assert relations between graphical entities and some primitive geometric property that can be derived from geometrical procedures. In order to illustrate more fully the procedural aspect of non-boolean valued operations, we shall now consider those with rank *bool line line, dot*. Consider again Fig. 1. The vector A is defined by the position vectors u_0 and u_1, and the vectorial equation $A = u_1 - u_0$ holds. We have a similar equation for vector B. The position vector p of the intersection point between two arbitrarily defined vectors $u_1 - u_0$ and $v_1 - v_0$ in the space is defined in terms of the following equations, where t_1 and t_2 are scalar parameters:

(47) $p = u_0 + t_1 * (u_1 - u_0)$
$p = v_0 + t_2 * (v_1 - v_0)$

We classify the scalar parameter values t_1 and t_2 into five cases: $t_i \leq 0$, $t_i = 0$, $0 < t_i < 1$, $t_i = 1$ and $t_i > 1$. Within this classification there are 25 cases of parameter pairs, as shown on Fig. 2. Every intersection case is computed by an operation of the form *int_xy*(ϕ, α, β), where α and β are two arbitrary vectors in the space, and ϕ is a boolean condition. The name of the operator for every 'overt' intersection case is of the form *int_xy* where x and y are symbols in the set $\{o, m, e\}$. These symbols stand for *origin, middle* and *end* respectively. The interpretation of these operators names is as follows: if x is o, the intersection point p between α and β occurs in the *o*rigin of α; if x is m, the intersection point p occurs in the *m*iddle of α; and if this symbol is e then p occurs at the *e*nd of α. The code symbol y relates the intersection point p with the vector β in a similar manner. The interpretation of operator *int_ww* covers the 16 cases in the outer-ring in Fig. 2. This set of cases is considered in the language for the definition of construction lines. In Fig. 2, the α vector is horizontally oriented, and the β vector is vertically oriented.

Now we consider the relation between the algebraic and algorithmic interpretations of **D** for operator symbols of rank *bool line line, dot*. As was mentioned, the operation *int_mm* denotes a function whose value is the intersection point of a 'cross intersection' between two vectors. The value of the boolean argument ϕ in *int_mm*(ϕ, α, β) is *true* if the pair ($0 < t_1 < 1$ and $0 < t_2 < 1$) in Fig. 1 is selected when α and β are compared. Then, *int_mm*(*true*, α, β) $= p$, where p is a 'normal element' in the carrier of terms of sort *dot*. However, if the value of the parameters pair is one among the other 24 possible cases, then *int_mm*(*false*, α, β) $= e_{<dot, i>}$.

The boolean argument in *int_em*(ϕ, α, β) *is true* if there is a *joint in t* intersection between α and β. The corresponding case in Fig. 2 is ($t_1 = 1, 0 < t_2 < 1$). When both parameter values are either 0 or 1 a joint of two vectors in their extreme points is determined. The operation symbols in this class are *int_oo, int_oe, int_eo, int_ee*. When one or both of the parameters take a value in either $t < 0$ or $t > 1$ there is no actual intersection, but if one or both vectors were projected, an intersection would occur. These cases are subsumed under the operator symbol *int_ww*. We also refer to this class of intersections as 'conditional intersections'.

Operators of rank *bool polygon polygon, polygon*, namely *union_of, intersection_of* and *difference_between*, are interpreted as a topological operation between polygons. The inclusion of polygon comparison operations allows the definition of complex regions of the space by composition of simple polygons. These operators can be defined in **D** if polygons are defined as regular sets of dots, given that set comparisons between two arbitrary polygons result in normal elements of sort *polygon* [13]. Algorithms for comparing two arbitrary polygons can be efficiently implemented [5, 14]. The output of these operations is illustrated in Fig. 3.

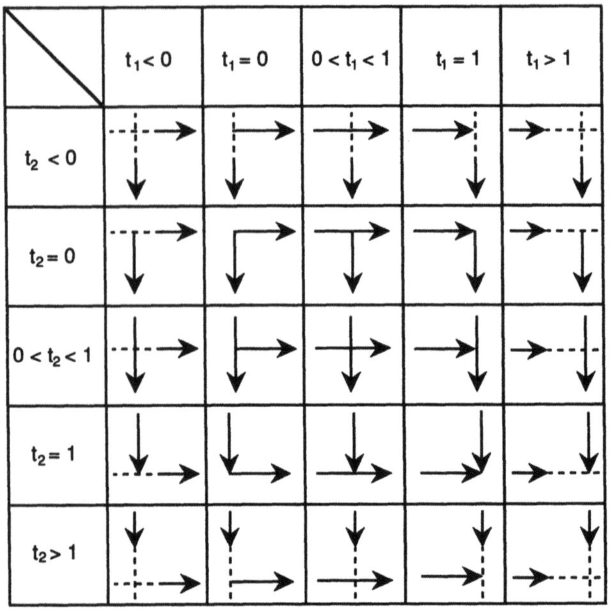

Fig. 2. The 16 cases for the definition of construction lines

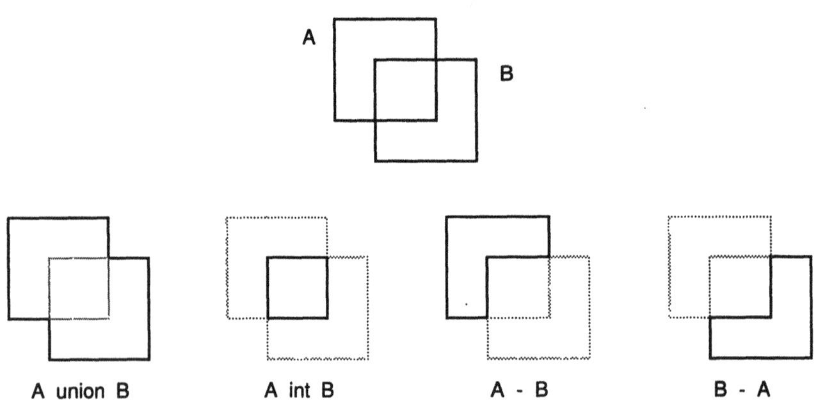

Fig. 3. The output of polygon comparison operations

The algorithmic interpretation of the other operators in **D** is computed by standard geometric algorithms and will not be discussed further here; see, for example [9, 11].

3. Intensionality in L_d

The language $\mathbf{L_d}$ is intended to express design knowledge in a very simple design domain. Design can be thought of as an activity in which a design object and its properties are identified. Design statements express propositions about objects that do not yet exist in the actual world. However, we can say that a design object is a part of some possible world. For instance, I can say that I am going to design *my house* even if there is no concrete thing in the world that is 'my house'. But there might be a future time in this world in which such a house has been built, or there might be a world that I can imagine in which such a house is already there. This name can be, of course, a graphical representation: the drawing of such a house.

In the design process we discover a set of properties and relations that the design object must have and preserve. However, all of these properties are contingent and varying some, or even all of them, does not destroy the identity of the object as long as there is the intention of designing such an object. We can even consider varying its name, but a naming accident is unlikely to change the identity of the object itself. The same can be said of drawings: there might be many graphical expressions that refer to my house and if I am designing this object I must be able to tell whether or not the drawing –or more generally, the representation– that is produced at some particular design state refers to my house. Of course, there might be several possible representations, but in many situations finding just one is enough. However, in order to make such an identification we need a criterion of identity. This poses a rather puzzling question: how can we identify something that does not yet exist in the actual world? Or rather how can we identify something that is undergoing a process of change?

At the beginning of interactive computer graphics, Sutherland advanced a criterion for the identification of design objects that are represented through drawings in the seminal Sketchpad program: [12]

> Construction of a drawing with Sketchpad is *itself* a model of the design process. The locations of the points and lines of the drawing model the variables of a design, and the geometrical constraints applied to the points and lines of the drawing model the design constraints which limit the values of design variables. The ability of Sketchpad to satisfy the geometric constraints applied to the parts of a drawing models the ability of a good designer to satisfy all the design conditions imposed by the limitations of his materials, cost, etc. In fact, since designers in many fields produce nothing themselves but a drawing of a part, design conditions might well be thought of as applying to the drawing of a part rather to the part itself. When such design conditions are added to Sketchpad's vocabulary of constraints, the computer will be able to assist a user not only in arriving at a nice looking drawing, but also in arriving at a sound design.

In GRAFLOG, we explore the same issues from a semantic approach. For this, we introduce the notion of *intension* of a graphical representation.

3.1. Extension and intension of graphical objects

The relation of equality poses many interesting questions about the identity of an object. To understand a statement of the form $\alpha = \beta$, we have to know what object is designated by the expressions α and β. To understand the relation of equality, we have to consider as well 'the manner' or 'mode of presentation' in which an expression designates an object. A statement of the form $\alpha = \beta$ suggests that there are two different modes for designating the same thing. For instance, expressions *3* and *2 + 1* designate the same number. Geometrical objects can be designated in different manners too. Frege gives the following example [3]

> Let *a, b, c* be the lines connecting the vertices of a triangle with the midpoints of the opposite sides. The point of intersection of *a* and *b* is then the same as the point of intersection of *b* and *c*. So we have different designations for the same point, and these names ("point of intersection of *a* and *b*", "point of intersection of *b* and *c*") likewise indicate the mode of presentation; and hence the statement contains actual knowledge.

Frege's triangle is illustrated in Fig. 4.a.

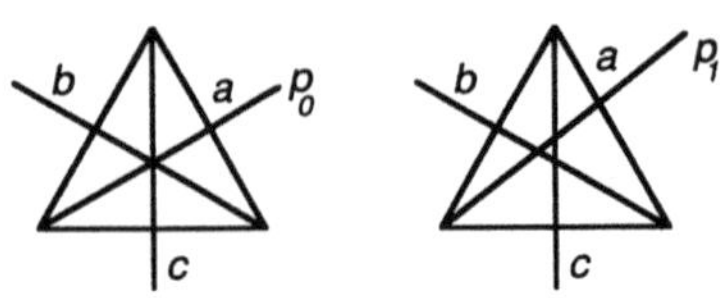

a) b)

Fig. 4. Frege's triangle

In relation to the triangle in Fig. 4.a, the statements "point of intersection of *a* and *b*" and "point of intersection of *b* and *c*" present the same object in two different manners. These two expressions refer to the same object, but they express different concepts and have a different informative content. According to Frege, they express a different **sense**.

Now, consider the interpretation of a statement whose form is $\alpha = \alpha$. It holds *a priori* given that the only thing we need for understanding the truth that it expresses is knowledge of the language. The statement "point of intersection of *a* and *b* = point of intersection of *a* and *b*" expresses an analytic truth. The truth of this statement is granted just by the form of the expression, and we do not have to think of the contingent properties of the lines *a* and *b*, such as their length and position, because there is no drawing in which this statement becomes false. Through $\mathbf{L_d}$ we can generalise this definition for two arbitrary lines in $\mathbf{L_d}$ as follows:

(48) $\Box\forall \mathbf{x}, \mathbf{y}_{\mathbf{line}}[\mathbf{int\_mm(x, y) = int\_mm(x, y)}]$

Note that int_mm(x, y) can denote a normal element of sort dot or the error element $e_{<dot,\, i>}$, but in any case (48) holds for every state *i* in *I* in any model for $\mathbf{L_d}$.

Consider now a statement of the form $\alpha = \beta$. For instance, "point of intersection of *a* and *b* = point of intersection of *b* and *c*." We can imagine some situations in which this statement is true, as is the case in Frege's triangle, but there are other graphical contexts in which it is false, for instance, the state of affairs illustrated in Fig. 4.b. Here, the position p_0 of one of the extreme points of the line *a* has been changed to p_1 and some of the contingent properties of this line –like length and orientation– have been altered. Then, the statement "point of intersection of *a* and *b* = point of intersection of *b* and *c*" becomes false because the intersections of *a* and *b* and *a* and *c* are different objects of sort dot.

So-called extensional logical systems with equality obey *Leibnitz's Law*. This states that if a statement of the form $\alpha = \beta$ is true, and if a statement ϕ containing α is true, then the result of replacing any occurrence of α in ϕ by β is also true. Consider, for instance, the following expression:

(49) **on(int_mm(a, b), a)**

where a and b are the lines in Frege's triangle in Fig. 4.a. This expression asserts that the point of intersection between the lines a and b is on the line a. The expression int_mm(a, b) is a term of sort dot that refers to the intersection of a and b and (49) is true in relation to the drawing in Fig. 4.a because this point is also on line a. If we substitute int_mm(a, b) by another term that has the same referent, say int_mm(b, c), the referent of the composite expression –its truth value– should not be altered. In fact, (50) is true in relation to the drawing in Fig. 4.a as well.

(50) **on(int_mm(b, c), a)**

An interesting point that was originally noted by Frege is that such a kind of substitution is not allowed in the context of modal expressions such as *necessarily*. This corresponds with the necessity operator □ of $\mathbf{L_d}$. Note that the truth value of expression (51) is true,

(51) **□on(int_mm(a, b), a)**

but the truth value of (52) is false.

(52) **□on(int_mm(b, c), a)**

Expression (52) is false because there are situations such as the case in Fig. 4.b where the intersection of b and c is not on the line a.

Failure of this kind of substitution implies that the extension of a complex expression at some given state of affairs is not necessarily a function of the extension of its constituent parts in such a state. Frege noticed this problem in relation to the meaning of natural language expressions, and advanced a hypothesis that has become fundamental for modern semantic studies. According to Frege, the semantic value, or denotation of an expression that appears within the context of an 'opaque operator'[2] like modal expressions, is not the normal denotation of such a term, but rather its sense. In the semantic interpretation of $\mathbf{L_d}$ we have adopted Frege's approach in an explicit way. But before explaining how this assumption has been implemented we have to be more specific on the notion of sense or intension of expressions of $\mathbf{L_d}$.

Following Montague and others [2] we define the intension of an expression as a function from states to extensions. In $\mathbf{L_d}$, the intension of an expression of sort s is a function whose domain is the set I of states and whose range is contained in the set D_s. The underlying idea is this: if the intension of an expression is known, its referent –or extension– can be identified at every state of affairs in I. Next, we illustrate this notion with the help of some simple examples.

Assume that Figs. 4.a and 4.b are respectively associated with the indices i_0 and i_1 in the set of states I of some design process.

The intension of an individual α is a constant function that assigns the same object to α for every interpretation state. For instance, the intension of line a is the function $\{<i_0, A>, <i_1, A>\}$. Note that we use the same symbol a as a name in $\mathbf{L_d}$ and A as member of the carriers of sort line. However, these two objects should not be confused. A name is a linguistic object, and the range of the function is a set theoretical object. Alternatively, we can think of the linguistic name a as the graphical symbol itself.

---

[2] These kinds of context are known as *referentially opaque constructions*. See, for instance, Dowty et al. pp.143 [2].

The intension of an expression ϕ which express geometrical properties of graphical objects is a function that maps every design state i to the value of ϕ in i. For instance, the position of the end-extreme of a in Fig. 4.a is p_0,[3] but the same value of the same expression in Fig. 4.b. is p_1. Accordingly, the intension of the expression position_of(end_of(a)) is the function: $\{<i_0, p_0>, <i_1, p_1>\}$.

The intension of a term of sort bool –a formula– is a proposition: a function mapping every design state to the truth value of such a formula in that state. For instance, the proposition that on(int_mm(b, c), a) expresses is the function $\{<i_0, 1>, <i_1, 0>\}$.

A question that is relevant for our enterprise is how the intension of an expression can be known. In modern formal semantic studies, the interpretation function relating the expressions of a language with their corresponding referents in every state is given by definition, and the epistemological question of how such a function can be known is not usually faced directly. In GRAFLOG, the knowledge of the intension of a graphical expression is embedded in the geometrical algorithms that compute its referent. For instance, the position of the point that is referred to by the term int_mm(a, c) in Figs. 4.a –an object of sort real_pair– can be discovered by computing the value of the expression position_of(int_mm(a, c)) at that state. The value of this expression is either a normal element of sort real_pair or the error element of this sort. The referent of such an expression can be determined in other interpretation states in the same manner. If we know the geometrical algorithms that compute the functions named by the operator symbols of **D** then we can come to know the intension of every graphical expression of **D** and this is in fact the case.

We can now explain how the principle of compositionality is implemented in the interpretation of $\mathbf{L_d}$. The meaning of constant and predicates names is defined by the interpretation function F. The extensions of individual constants and predicates depend on the current state of the knowledge-base. However, the extensions of graphical terms of D are computed through geometry. The extension of a complex but non-modal expression is determined by applying the function denoted by the operator term to the extensions of its arguments, as defined by the semantic rules in the interpretation of $\mathbf{L_d}$. In this case, the extension of a complex expression is a function of the extensions of its constituent parts at the current interpretation state. However, the semantic value of expressions of the form $\Box\phi$ is a function of the intension of ϕ. In the context of the modal operator, ϕ has not its customary denotation, but rather denotes the proposition that it expresses. We require for $\Box\phi$ to be true that the semantic value of ϕ is true at every index i in I.

Now, we give some formulas of $\mathbf{L_d}$ that hold for every interpretation state.[4]

(53) **□∀x, y int_mm(x, y) ≡ cross_at(x, y).**

(54) **□∀x, y, z [x = int_om(y, z) ∨ x = int_em(y, z) ∨ x = int_mo(y, z) ∨ x = int_me(y, z)]**
≡ x = t_joins_at(y, z).

---

[3] Such a coordinate pair can be referred to by means of operator symbols (7) and (11) of the structure **D**.

[4] Assume that quantifiers and variables in these formulas are of the appropriate sorts. We also omit the boolean argument for the error condition of each operator symbol for clarity.

(55) $\Box\forall x, y, z\,[x = \mathrm{int\_oo}(y, z) \vee x = \mathrm{int\_oe}(y, z) \vee x = \mathrm{int\_eo}(y, z) \vee x = \mathrm{int\_ee}(y, z)]$
$\equiv x = \mathrm{e\_join\_at}(y, z).$

(56) $\Box\forall x, y, z\,[x = \mathrm{e\_join\_at}(y, z) \vee x = \mathrm{t\_joins\_at}(y, z)] \equiv x = \mathrm{join\_at}(y, z).$

(57) $\Box\forall x, y, z\,[x = \mathrm{cross\_at}(y, z) \vee x = \mathrm{join\_at}(y, z) \vee x = \mathrm{int\_ww}(y, z)]$
$\equiv x = \mathrm{intersect\_at}(y, z).$

These formulas define a hierarchical categorisation on the 25 intersection cases in Fig. 2. For instance, the term e_join_at denotes any of the four modes in which an 'extreme joint' between two vectors can occur. The term t_join_at stands for the four cases of 'joins in t'. The term joint_at stands for any kind of join, and the term intersects_at denotes any 'actual' or 'possible' intersection between two lines. The term cross_at is just a more intuitive term to refer to the intersection int_mm.

Terms such as cross_at,joint_at etc., can be given a natural language translation, and can be used to talk about graphical relations. In the context of a graphical and linguistic conversation, it might be more natural to say, for instance, *This is the join of these walls* rather than *This is where the end of this wall joins the origin of this wall.* Although the latter expression is more informative, the former is general and abstracts over low-level geometrical and topological considerations. We can imagine different contexts in which it is better to utter one rather than the other. In a conversation between an architect and his customer, the former expression might be natural; but for a designer involved in the definition of the relations that must hold when a drawing is modified, say in a CAD system, the latter expression might be most appropriate. We include both kind of terms for referring to drawings at different levels of abstraction as a means of adding expressivity to our representational environment.

So far, we have explained in detail the notions of extension and intension of a drawing represented through the language $\mathbf{L_d}$. Next, we show the utility of these distinctions for defining a criterion of identity for graphical objects that undergo a process of change.

3.2. The identity of graphical objects in $\mathbf{L_d}$

One fundamental question that is not often raised in the definition of so-called design systems is how the identity of a graphical object is determined. In many traditional approaches design is considered an exploratory task in which a sequence of states of a design process are visited, and eventually, a design object is 'recognised' by the human designer. However, if something is identified at all it is because there is a criterion of identity. This criterion might be implicit; nevertheless, it has to be there. Here, we define a convention for identifying drawings when the geometrical properties of their constituent symbols are altered. Of course, this criterion is conventional and the design task on which we are about to embark is extremely simple. However, the criterion is useful for the solution of some problems that traditional design systems have often suffered from.

In GRAFLOG, the linguistic interpretation of a graphical expression is stated by means of an ostensive definition [6, 8]. Now consider the following question: does a linguistic symbol that is introduced by ostension name the extension of the graphical symbol or rather its intension? For instance, if the expression *This is a wall* is uttered at the time a line in an architectural drawing is pointed out, is the term *a wall* naming the intension or rather the extension of such a line? In this case we might say that *a wall* is naming an extension, that is, the object that the line stands for; this graphical symbol has its 'customary denotation'. However, given that basic graphical symbols denote the same object in every design state a basic symbol expresses an intension, although it is a constant function. On the other hand, consider the situation in which a region of the space that is determined by a set of walls is selected at the time *This is a house* is typed in. Here, we could also say that *a house* and the graphical symbol denote an extension: the house itself. However, if such a drawing is altered in the course of a design process –for instance, by modifying the length of the walls– how can we tell which space partition corresponds to the object named by the term *a house* in a new graphical configuration? In other words, how can we tell whether after some modification the drawing still denotes the house or whether it is just a meaningless set of graphical patterns? We assume that in this situation the term *a house* names the intension of the graphical symbol. We can be more specific: the denotation of a graphical symbol that emerges in the context of a graphical context is not 'its customary denotation' but rather its intension.

In the implementation of GRAFLOG we distinguish two kinds of ostensive definitions. The first kind is used for giving names to context-independent symbols whose internal geometrical 'unity' is given beforehand. For instance, the walls of an architectural drawing. The second is used for giving names to complex graphical structures that emerge from the composition of more basic units. For instance, for naming a region of the space standing for a house in terms of a set of basic lines.

Consider as well that an expression of $\mathbf{L_d}$ might denote normal objects of its corresponding sort, or the error element of that sort. We adopt the following convention: if the denotation of a graphical expression is a normal element of its corresponding sort, its extension corresponds with the object named by the expression; however, if its denotation is the error element of its sort we consider that the identity of the referred object is not fully determined in such a state. We adopt the additional convention that a name in the linguistic domain that is given to a graphical symbol denotes if and only if the current graphical term denotes a normal element of its sort. In other words, if the graphical expression denotes the error element of its sort, its linguistic name does not denote.

Next, we show how the representational environment provided by $\mathbf{L_d}$ is implemented and used in GRAFLOG.

4. Interactive Definition of Design Objects

In this section we present the definition of graphical structures whose basic symbols are expressed in the course of standard graphical interaction, and whose interpretation is captured in the language $\mathbf{L_d}$. For the implementation, two representational structures are used. The linguistic information is represented as Prolog's facts and relations and the graphical information is stored in a set of geometrical clauses g_db.

Here, we consider a graphical menu with only one kind of symbol: the line. Symbols of sorts dot and polygon will be identified in terms of the geometrical and topological relations that are established between the basic lines. Now, suppose that the drawing in Fig. 5 is created by editing five lines, at the time the ostensive definition in (58) is typed in by the user.[5]

[5] For clarity, we show the order and orientation in which the lines are edited on the screen, but the labels and arrows are not part of the drawing.

(58) *These are walls.*

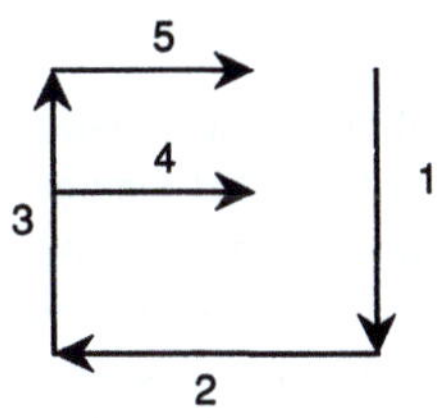

Fig. 5. Drawing of a house

The symbols introduced by this definition are independent of the graphical context, and they constitute the basic building blocks of the graphical composition. The graphical and linguistic knowledge that is expressed by this definition will update the knowledge-base and also the geometrical data-base g_db in an appropriate manner. In the linguistic domain, the following facts will be asserted, where the individual names of the lines, like wall_1, are arbitrary identifiers provided by the system:

(59) (1) **wall(wall_1).**
(2) **wall(wall_2).**
(3) **wall(wall_3).**
(4) **wall(wall_4).**
(5) **wall(wall_5).**

The geometrical data base is constituted by a set of clauses such that for each graphical object in the representation there is a clause of the form

(60) **g_db(name, type, description).**

Here, name is a constant identifying the graphical object, type is the name of some sort in $\mathbf{L_d}$, and description is a list of terms denoting the points of the space, the parameters, by which the graphical object is defined. We consider three kinds of descriptions: of basic lines, of polygons, and of construction lines.

The description of basic lines is stated as a list of two constant terms of sort real_pair which stand for the positions of the origin and end of the corresponding line.[6] These terms are of course computed by the system at the time the symbols are defined on the screen. The notation is as follows: p_{ij} is interpreted as point i of line j, where i is o for the origin and e for the end of the corresponding line. For instance, p_{o4} is interpreted as the position of the origin of the fourth line. Definition (58) is reflected in the graphical domain as follows:

(61) (1) **g_db(wall_1, line, [** p_{o1}, p_{e1}**]).**
(2) **g_db(wall_2, line, [** p_{o2}, p_{e2}**]).**
(3) **g_db(wall_3, line, [** p_{o3}, p_{e3}**]).**
(4) **g_db(wall_4, line, [** p_{o4}, p_{e4}**]).**
(5) **g_db(wall_5, line, [** p_{o5}, p_{e5}**]).**

[6] Objects of sort dot are not implemented directly, and dots are represented through their corresponding positions.

The identifier of a given object in g_db is also used as the name of the same individual in the domain knowledge-base. For instance, the object that has the property of being a wall in (59.1) is depicted on the screen in terms of the graphical information in (61.1).

Now, we come to the definition of the second kind of graphical symbol and its associated description list: objects of sort polygon. Polygons are used for representing entities that emerge in the context of overt lines. The interpretation of a polygon is stated through an ostensive definition. However, this kind of ostension differs from the previous one in that polygons are identified in terms of the basic symbols that are already drawn on the screen. This ostension does not introduce overt graphical patterns on the screen but rather establishes a relation between symbols, like the walls, that were introduced before. The linguistic name introduced by this kind of definition names not only a polygon in the actual definition state, but rather the function by which the polygon can be known in different graphical states. The expressions included in the description list of polygons are intensions rather than extensions.

Suppose that when the expression

(62) *This is a house*

is typed in, a polygon whose vertices are the dots identified in Fig. 6 is defined on the screen by the user. This identification consists in pointing out the dots in the order indicated by the subscripted values.

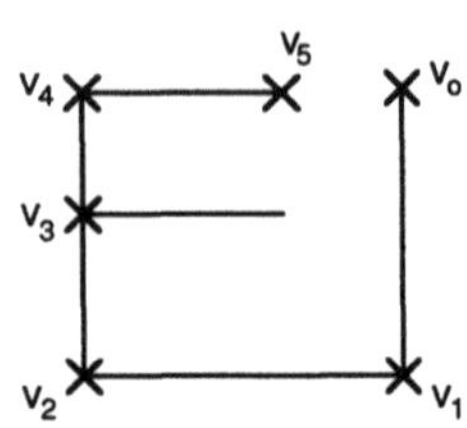

Fig. 6. Dots identifying the vertices of a polygon

The definition can be started by pointing at any of those vertices, but the order in which these are identified will be reflected in the polygon's representation in g_db; otherwise, a different object would be defined. In the definition of this polygon we have used a graphical cursor of sort dot. This is of course an implementational convention and other strategies could be considered; for instance, we could have pointed to the screen with a graphical cursor of sort polygon. In Fig. 6, the edges of the polygon are not explicitly drawn, but this is also a convention.

The linguistic and graphical expressions representing this object are asserted in the representational structures of GRAFLOG. The fact

(63) **house(house_1).**

is added in the knowledge-base. In the graphical domain, the graphical object is represented by a clause in g_db. The form of this clause is as follows:

(64) **g_db(house_1, polygon,**
[origin_of(wall_1, _),
e_join_at(wall_2, wall_1, _),
e_join_at(wall_3, wall_2, _),
t_join_at(wall_3, wall_4, _),
e_join_at(wall_5, wall_3, _),
end_of(wall_5, _)]).

Note that the polygon's description takes the form of a list, where each term is the translation into Prolog of a term of sort dot of $\mathbf{L_d}$ that denotes one of the polygon's vertex. These terms denote the origin or the end of a line, or the intersection of two lines.[7] Obviously, the translation from $\mathbf{L_d}$ in Prolog has to accommodate the fact that Prolog does not allow functions to be expressed directly. Thus, in place of the $\mathbf{L_d}$ operator $origin\_of_{line,\ dot}$ we use a Prolog relation origin_of/3 where the last argument-place stores a term denoting the dot which will be the value of the function. We implement other operator symbols of **D** in a similar way.

According to our criterion of identity for graphical objects, the linguistic designator house_1 names the intension of the graphical symbol that it indexes. Additionally, house_1 will denote at every state in which its associated polygon is a normal element sort polygon.

The identification of graphical objects in different states is constrained by the geometrical relations between the terms included in the description list of these objects. In order to denote, graphical symbols must satisfy the following conditions:

(65) If the sort of the symbol is line, the positions of the dots defining such a line must be distinct.

(66) If the sort of the symbol is polygon, the terms of sort dot in the description list define a simple closed path in which no edges intersect each other.

With these considerations, we can update the lines defining the house, as shown in Fig. 7.

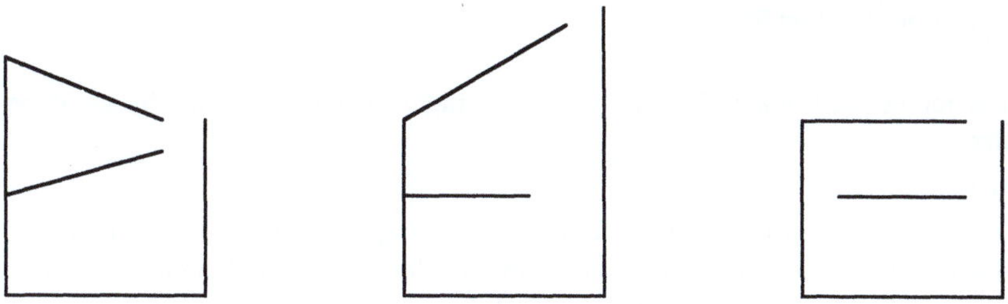

a) b) c)

Fig. 7. Three variations of 'house_1'

[7] The expressions representing the polygon's vertices are inferred to by GRAFLOG's interpreter in the course of the graphical interaction, as will be shown when the graphical parser for **D** is explained.

Under the current house's definition, Figs. 7.a and 7.b are permissible variations of 'house_1'. However, the drawing in Fig. 7.c cannot be identified as the same entity, because one topological specification in the original definition is not satisfied by such a drawing. The error condition can be expressed in $\mathbf{L_d}$ as follows: t_join_at(wall_3, wall_4) = $e_{<dot,\ i>}$ As a consequence, house_1 = $e_{<polygon,\ j>}$.

Here we can also appreciate why the name of a graphical symbol does not denote if the symbol is in an error state. If we are engaged in the task of designing house_1, it makes sense to say that the drawings in Figs. 7.a and 7.b are, according to definition (62), graphical representations of the house that we are designing, but it makes no sense to say the same of the drawing in Fig. 7.c.

Now, we come to the definition of the third kind of graphical object: construction lines. Construction lines are required for identifying space partitions that are not fully determined by the origin, the end or the intersection of overt lines. Consider the definition of a room of the house. This room is defined be typing in the following expression:

(67) *This is a room.*

Assume that expression (67) is supported by pointing to the marks in Fig. 8.a, as was shown above for the house.

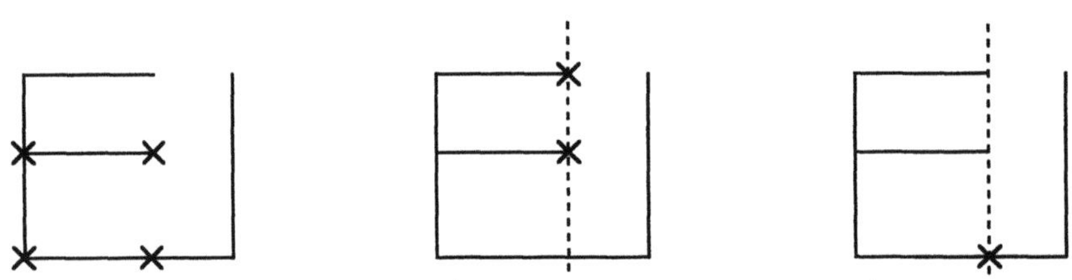

a) b) c)

Fig. 8. Definition of a construction line

The room is represented along the lines of other graphical objects named by common nouns, and the fact

(68) **room(room_1).**

is asserted. For the definition of the graphical object we require, however, a construction line. Three dots are identified as the origin or the end of a line standing for a wall; but one vertex (i.e. the one in the lower right-hand corner) cannot be fully determined just by virtue of the extremes and junctions of the walls.

Construction lines can be defined by selecting two points on the screen as shown in Fig. 8.b. This line is represented in the graphical domain as follows,

(69) **g_db(c_line_1, line, [end_of(wall_5, _), end_of(wall_4, _)]).**

The identifier c_line_1 is provided by the system, and the description list is constituted by the two terms of sort dot that refer to the points identified in Fig. 8.b. Note that this line 'emerges' from the graphical context in which it is defined. This is an important consideration because if the lines in terms of which the construction line is defined, namely wall_5 and wall_4, are altered, the construction line will be modified accordingly. However, the relations in the description list of the construction line will remain constant across changes. What varies is the extension of c_line_1, but its intension remains the same. In the same way that the description list of a polygon contains the intension of such a polygon, the description list of a construction line is an intensional description.

We consider a construction line to be the infinite projection of the vector that the terms in its description list define. For that reason, points identified by a construction line are referred to by the operation symbol intersect_at in (57), where one or both of the arguments of sort line of this operation symbol are construction lines themselves. The function that is named by intersect_at has as its value a normal element of the carrier of sort dot, unless the two lines that it takes as arguments are parallel or collineal.

Now we show how the construction line is used for the definition of the bottom-right vertex of the room. It corresponds with the intersection of c_line_1 and wall_2, as shown in Fig. 8.c. Such a dot is denoted by the term intersect_at(c_line_1, wall_2) in $\mathbf{L_d}$. The other three vertices are identified in terms of the basic lines, and the entry for the room in the geometrical representation is the following Prolog clause:

```
(70) g_db(room_1, polygon,
        [end_of(wall_4, _),
         intersect_at(c_line_1, wall_2, _),
         e_join_at(wall_3, wall_2, _),
         t_join_at(wall_3, wall_4, _)]).
```

Now we can appreciate that the bottom-right vertex of the room will be properly defined in different states of the graphical representation unless the two points defining the construction line become the same, or the construction line and wall_2 become parallel or collineal. In the case that either of these conditions occurs, not only the bottom-right vertex of the room becomes an error of sort dot, but the room itself becomes an error of sort polygon.

Before concluding this section it is worth summarising the relation between the abstract representational language $\mathbf{L_d}$ and the representational structures of GRAFLOG that are given in terms of Prolog clauses.

First, every graphical symbol represented by a clause g_db in the implementation corresponds to a basic constant in the abstract specification in $\mathbf{L_d}$. For instance, the individual denoted by the name wall_1 in (59.1) and (61.1) happens to have the property of being a wall, and happens to have a graphical realisation as a line as well. The points p_{o1} and p_{e1} in its description list in (61.1) are individuals of sort dot, and they are basic constants in the abstract representation language $\mathbf{L_d}$ as well. Note that the relation between formal structures in $\mathbf{L_d}$ and the implementation in GRAFLOG is captured by constructing the description list of complex graphical objects, like lines and polygons, out of terms of a more simple geometrical type, like dots.[8] The translation of expressions of $\mathbf{L_d}$ into Prolog is as follows:

[8] However, there is no an explicit counter part in $\mathbf{L_d}$ of the geometrical and topological relations between basic graphical objects and the more complex object that they construct. For instance, the relation part_of(p_{o1}, wall_1) is neither implied nor required in the representational system. The linguistic description of a drawing does not necessarily corresponds with the structure of such a drawing.

(71) Every clause of the knowledge-base is the translation into Prolog of an expression of sort bool in $\mathbf{L_d}$.

(72) Every clause g_db in the geometrical data-base corresponds to a basic constant in $\mathbf{L_d}$ that names an object that has both abstract and graphical properties.

(73) Every term in the description list of graphical objects of types line and polygon in g_db is the translation into Prolog of a term of sort dot.

An important distinction that might require further clarification is which properties of the representational system $\mathbf{L_d}$ and its Prolog implementation remain invariant across different interpretation states. That is to say, what properties of $\mathbf{L_d}$ hold for every possible model, on the one hand, and which kind of information is contingent and depends on the current information state of the system, on the other.

In the definition of formal languages, two kind of constants normally distinguished: logical and non-logical. The semantic interpretation of logical constants is defined by an explicit function that holds for every model of the language. For instance, the semantics of *and*, *or*, etc. are given by their corresponding truth-tables. On the other hand, the interpretation of non-logical constants –individual constants, predicates and relation names– is stated by a valuation function *F* of the model for the language. This function changes from model to model.

In our theory, the interpretation of logical constants of $\mathbf{L_d}$ was defined when the algebraic interpretation of the language was presented. The interpretation of truth-functors, as '$\wedge$', '$\vee$', '$\rightarrow$', etc. was stated in a standard way. Additionally, the interpretation of operation of symbols of **D** was stated once and for all in clause (38). The function *F*, on the other hand, is dynamically defined in every interactive session and depends on the current state of the information of the system.

In the interpreter of GRAFLOG, each logical constant of $\mathbf{L_d}$ is represented by a primitive functor of Prolog. For instance, the implication '$\rightarrow$' is translated as ':=', and the negation symbol '$\neg$' is translated as Prolog's functor *not*. Of course, the meanings of these operator symbols are not completely equivalent, and for the implementation some concessions have to be made. As is well known, there is an important difference between logical and Prolog negation [10]. However for the purpose of the implementation we adopt the convention that these meanings are equivalent. Additionally, there is in GRAFLOG's interpreter a primitive functor that corresponds to each operation symbol of **D**. For instance, end_of, intersect_at and $\cup$ are primitive GRAFLOG's functors. When the interpretation of an expression that is formed by these operation symbols is required, its associated geometrical algorithm computes the corresponding function's value. These algorithms are given beforehand to the system, and the function that they compute is the same in every model for the language.

Next, we come to the kind of information that is contingent and depends of the state of the interactive session. In fact, every state of the knowledge-base and g_db implicitly defines a particular model. For instance, the constant wall_1 is graphically realised as a line, and both of these symbols refer to an abstract object *wall_1* that is only in our minds. However, this association is model dependent. If the line is deleted in the course of the graphical interaction, the association vanishes, and the definition of *F* changes accordingly. *wall_1* is just an abstraction that we know when we interpret the expressions of the language. In a similar way, the predicate wall, for instance, is interpreted as the set $\{wall\_1, wall\_2, wall\_3, wall\_4, wall\_5\}$. The interpretation function maps such a predicate name to the set of objects that happen to have the property that wall names. It is worth emphasising that the objects in this set, the set itself, and the map between predicate name and set are also abstractions in our mind.

Here, we conclude the illustration of the graphical and linguistic definition of 2-dimensional wire-frame diagrams in this version of GRAFLOG. Next, we show how this kind of drawings can be queried in the course of graphical interaction.

4.0.1. Deictic questions in graphics interaction. The user can address queries about the meaning of the drawing. Suppose, to continue our example, the user picks out one of the locations indicated by a cross in Figs. 9.a, 9.b or 9.c, and then types,

(74) *What is this?*

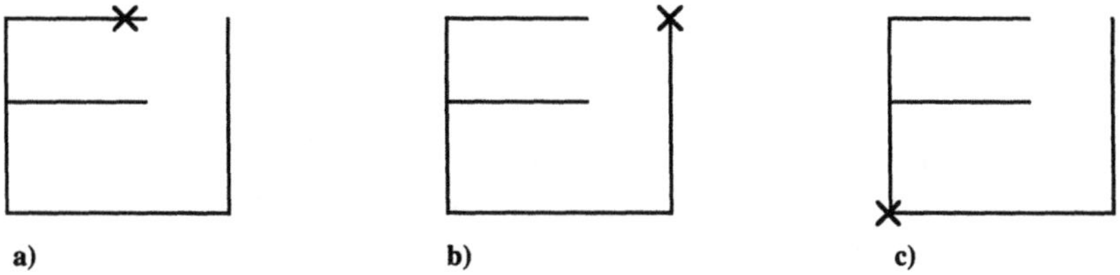

Fig. 9. Locations picked out by the user

The answers produced by the system depend on the drawing and the position of the pointing device. Thus, for each of the locations selected above, the responses for the corresponding questions will be

(75) (a) *A wall.*
(b) *The origin of this wall.*
(c) *The join of these walls.*

The linguistic answers are supported by graphical feedback as shown in Fig. 10. That is, the deixis required by the terms *this wall* and *these walls* is supplied graphically by highlighting the relevant components of the drawing on the screen. This 'highlighting' is indicated by the dotted lines below.

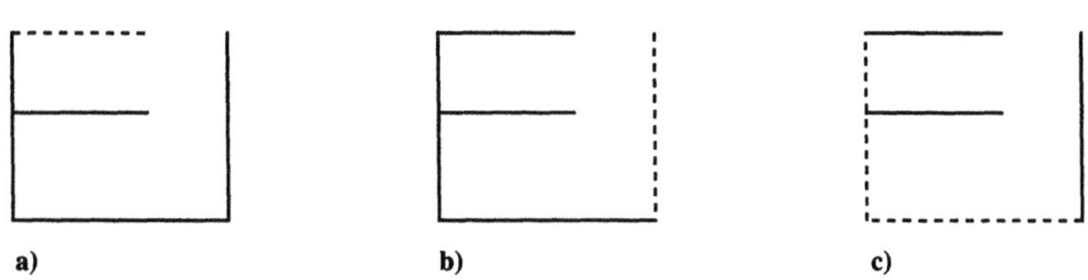

Fig. 10. Graphical feedback by highlighting the relevant components

The expressions *origin of* and *end of* in the natural language answers in (75.a) and (75.b) correspond to the operators origin_of and end_of in $\mathbf{L_d}$. The words *a, the, this* and *these* are 'function' words that determine quantification and deixis in these referring expressions [8]. We can ask as well for the identity of objects represented by symbols of sort polygon. If the location in Fig. 11.a is pointed out, and the question

(76) *What is this?*

is asked, the answer is *A house.*

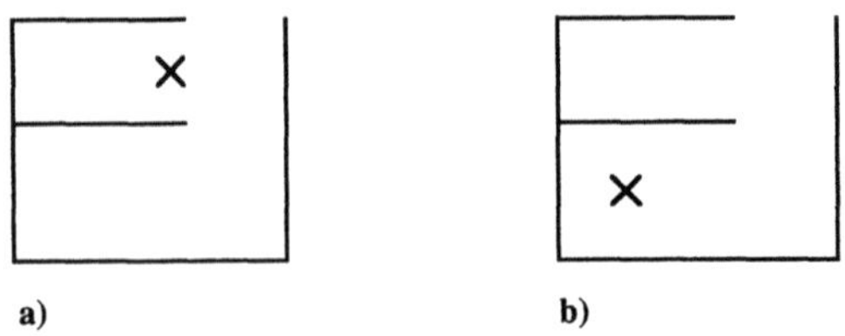

Fig. 11. Locations identifying a polygon

However, if the same question is asked when the location in Fig. 11.b is selected, there is a referential ambiguity. The object that is pointed out might be either the house or the room. Here, the system can produce an answer in terms of a geometrical default value and the answer could be, for instance, *a room.* However, the issue of answering these kind of question is rather complex, and further research is required in this matter [8].

5. The Parsing of Wire-frame Drawings in $\mathbf{L_d}$

The goal of the graphical parsing procedure is to find out the interpretation in the structure **D** of the graphical entities that are pointed out on the screen in each individual pointing action. Graphical objects are identified by a graphical cursor. Given that there are three sorts of objects that have a graphical representation, we can use three kinds of cursors for identifying objects of their corresponding sorts. We consider in detail the use of a graphical cursor of sort dot, and sketch the utility of cursors of sorts line and polygon.

In general, there are several terms that denote the same point at some state of the graphics interaction. For instance, the dot v_2 in Fig. 6 was identified by the term e_joint_at(wall_3, wall_2) in the corresponding entry in the description list of house_1. However, other terms, like int_eo(wall_3, wall_2) or int_oe(wall_2, wall_3) could also be considered. Although all of these terms denote the same object at the time in which the house is identified, they have a different informative value and express a different proposition. Here, we have to choose which term is the one that best expresses the knowledge intended by the user. We partition the term selection task in two stages. First, we have to find out the set of all terms that denote the mark of the pointing device in an individual pointing action. Second, we define a criterion for selecting the term that has to be included in the description list of the identified object. We discuss these in turn.

An expression of the graphical language $\mathbf{L_d}$ can be thought of as a partial description of a drawing. That is, an expression of the form p = position_of(origin_of(α)), where p names a coordinate point and α is the description of a line, is construed as a statement which is true or false of a given two-dimensional wire-frame drawing Γ. In this case, the drawing plays the role of being a model for a set of expressions of $\mathbf{L_d}$, and we would need to develop a recursive account of expressions being true in such a model.

Alternatively, we can think of a drawing Γ as being a collection of well-formed terms of $\mathbf{L_d}$, closed under some rules of inference (e.g. *modus ponens*). In this case, Γ is construed as a theory in $\mathbf{L_d}$. Here, we need to develop an account of which logical and non-logical axioms would be required for finding out the set of equations that hold for a drawing in terms of the basic graphical facts. We require as well some proof procedures for this theory. In the current approach we explore the second alternative.

Thus, we think of a drawing as being a set Γ of terms and expressions of the language $\mathbf{L_d}$. We also need to encode information about the pointing device, the graphical cursor mark, at any given time. We use the contextual information provided by the set I of states of an interactive session as a way to encode the deictic aspect of mark. In standard 'indexical semantics', an expression like *here* is evaluated in a particular context of use i, which then supplies an appropriate referent, say $\mathbf{here}_i$. By analogy, we treat mark as a distinguished individual constant of sort dot in $\mathbf{L_d}$, such that for any graphical state i, $F(\text{mark})(i) = m_i$ where m_i is the dot selected by the graphical cursor in that state i.

We also parameterise Γ in two respects. First, since objects might disappear or be created as we pass from state to state, we need to treat a drawing as a set $\Gamma(i)$ of graphical objects at a given state i. Second, we distinguish the subsets $\Gamma_s(i)$ which belong to a particular sort s. Thus, for example, $\Gamma_{line}(i)$ is the set of terms in Γ at state i which have the sort line. In our example,

(77) $\Gamma_{\mathbf{line}}$ **= {wall_1, wall_2, wall_3, wall_4, wall_5, c_line_1}.**

Given the following set of inference rules, we will be able to derive various theorems from $\Gamma(i)$. We write

(78) $\Gamma(\mathbf{I}) \vdash \phi$

to specify that the expression ϕ is derivable, or can be deduced, from the theory $\Gamma(i)$.

If t_1, t_2 and t_3 are terms of $\mathbf{L_d}$, (79.1), (79.2) and (79.3) are valid inference rules:[9]

(79) (1) $\Gamma(\mathbf{i}), \mathbf{t_1 = t_2} \vdash \mathbf{f(t_1) = f(t_2)}$

(2) $\Gamma(\mathbf{i}), \mathbf{t_1 = t_2} \vdash \mathbf{t_2 = t_1}$

(3) $\Gamma(\mathbf{i}), \mathbf{t_1 = t_2, t_2 = t_3} \vdash \mathbf{t_1 = t_3}$

For example, suppose that the drawing represented by the set $\Gamma(i)$ includes {wall_1, wall_2, end_of(wall_1) = e_join_at(wall_1, wall_2), mark = position_of(end_of(wall_1))}. Then we have

(80) $\Gamma(\mathbf{I}) \vdash$ **mark = position_of(e_join_at(wall_1, wall_2))**

Let $MARK_{dot}(i)$ be the set $\{\alpha \text{ in } \Gamma_{dot}(i) \mid \Gamma(i) \vdash \text{mark} = \text{position\_of}(\alpha)\}$. This is, every expression α that denotes the same position as mark in the state i belongs to $MARK_{dot}(i)$. Next, we give an inductive definition of how to determine the set of terms in $MARK_{dot}(i)$.

---

[9] (79.1) is not defined if ϕ is opaque.

(81) For every α and β in $\Gamma_{line}(i)$ which are not parallel or collineal, if $\Gamma(i) \vdash$ mark = position_of(int(α, β)) then int(α, β)) is in $MARK_{dot}(i)$, where int is an intersection operator in clause (12) of the definition of **D**.

(82) For every α in $\Gamma_{line}(i)$ if $\Gamma(i) \vdash$ mark = position_of($\beta(\alpha)$) then $\beta(\alpha)$) is in $MARK_{dot}(i)$, where β is either end_of or origin_of.

(83) Nothing is in $MARK_{dot}(i)$ except as stipulated by clauses (81) and (82).

Now we come to the identification of lines and polygons by a cursor of sort dot. We define the sets $Selected_{lines}(i)$ and $Selected_{polygon}(i)$ of lines and polygons respectively that are identified by a cursor of sort dot at state *i*. The terms included in these sets are specified in the following inductive definitions:

(84) $Selected_{line}(i) = \{\alpha \in \Gamma_{line} \mid \text{on(mark, } \alpha)\}$.

(85) $Selected_{polygon}(i) = \{\alpha \in \Gamma_{line} \mid \text{in(mark, } \alpha)\}$.

In this section, the mark of the pointing device has been defined as a constant of sort dot. As was mentioned, a more flexible scheme can be defined by considering graphical pointing devices, graphical cursors, of sorts line and polygon. Such a functionality would allow not only the identification of basic lines and polygons but also of lines and polygons which emerge in drawings in terms of the overt symbols and the composition rules of **D**. The particular interest in our design domain context is the identification of polygons that are produced by the polygon comparison operations; that is to say, by terms produced by operators of rank bool polygon polygon, polygon in **D**. Such a device would allow us to impose linguistic interpretations by ostension on complex structures that emerge in drawings in a very direct manner. Suppose, for instance, that polygons *A* and *B* in Fig. 3 are interpreted as *linguistics* and *programming* respectively. Then we could point to the intersection between these two polygons with a pointing device of sort polygon –a graphical cursor for polygon definition– when the ostensive definition *This is science* is typed in. Now, suppose that one of the basic polygons is modified. The interpretation of the intersection, *science*, would vary accordingly. The geometrical clause for *science* is as follows:

(86) **g_db(science, polygon, [intersection_of(linguistics, programming, _)])**

where linguistics and programming are symbols of sort polygon.

The implementation of graphical cursors of sort line and polygon would allow the definition of the sets $MARK_{line}(i)$ and $MARK_{polygon}(i)$ along the lines of the set $MARK_{dot}(i)$ above. These facilities are useful because not only dots, but also lines and polygons can be referred to by a number of terms of $\mathbf{L_d}$ at any interaction state *i*.

Now, we come to the selection of the particular term in $MARK_{dot}(i)$ that is chosen by the GRAFLOG interpreter in the graphical parsing procedure. Note that the terms included in the description list of a polygon are not the basic operator terms of **D** which compute the actual intersection point, but rather they are terms produced by one of the intersection naming axioms which subsumes a family of intersection cases. The operators e_join_at, t_joint_at or intersect_at reflect better the knowledge about a graphical relation that a user might have at the definition state. Terms build up with these operators are less informative than the operator symbols that name the topological relations that actually hold between the vectors involved, but that information is hidden in normal human communication as well. When a human user points to an overt dot on a drawing, he or she might not be aware of the underlying geometrical information that is available through the representational system. Selecting these terms underspecifies the definition of the graphical object, but leaves a greater space for modification of drawings. If the more informative terms were included in the description list of a graphical object, the possible alterations for a drawing might be artificially restricted.

For the definition of the selection procedure, the terms that are considered for identifying a dot are ranked according their discriminatory value in relation to the other expressions that could refer to the same dot. For instance, both of the terms e_joint_at and t_joint_at cover four basic modes of intersection; however, the e_joint_at case marks not only the mode but also the location of the dot at which the intersection takes place. Accordingly, e_joint_at has a higher priority value than the term t_joins_at. We also take the conventional decision that origin and end points of vectors havê a larger priority of selection than the 16 intersection cases in Fig. 3 that are covered under the term int_ww. The operator term int_ww is not directly used, and it is subsumed in the intersection term intersect_at for the definition of construction lines. Terms that are included in the description list of a graphical object in g_db are selected according to the following priorities: cross_at = 5, e_joint_at = 4, t_join_at = 3, origin_of = 2, end_of = 2, intersect_at = 1, and int_xx = 0.

Now, the term of $MARK_{dot}(i)$ that is included in the description list of a graphical object is selected as follows:

(87) Among the terms cross_at, e_join_at, t_join_at, origin_of, end_of, intersect_at or int_xx select the term with a highest priority value. If there are several terms with the same priority, such that there are no terms with a higher priority, select any one of them.

(88) If $MARK_{dot}(i)$ is an empty set, no intersection is identified.

For instance, for selecting the term that refers to the point v_0 in Fig. 6 and is included in the description list of *house*, the set $MARK_{dot}(i)$ is

(89) **{origin_of(wall_1), int_ww(wall_5, wall_1), int_ww(wall_1, wall_5)}**

According to the selection procedure, the term included in the description list of house_1 is origin_of(wall_1).

6. Causal Relations and the Interpretation of State Transitions

Intersections of construction lines are useful for modelling causal relations in the process of modifying a drawing. In fact, these points provide a reference that is common to the drawings before and after a modification takes place. More generally, an intersection point –between overt and construction lines– is a kind of 'pivot' over which transformation functions can be defined.

Let us define a geometrical constraint that a drawing must satisfy in the course of a design task. We define as well a transformation function that reinforces such a constraint on a drawing whenever is required. Suppose that in the definition of the house, we define an additional construction line, namely c_line_2,[10] and it extends from the origin of wall_5 to the origin wall_1,

[10] c_line_1 was introduced in the definition of the room.

the right wall. Now, suppose that we impose the following two constraints upon the drawing that hold for every design state:

(90) **collineal(wall_5, c_line_2).**

(91) ∃**x.x = join_at(wall_5, c_line_1).**

The interpretation of these formulas is as follows: for every design state, (90) requires that the front wall, wall_5, has the same direction as the construction line c_line_2 and (91) stipulates that an extreme of wall_5 joins the other construction line c_line_1. The references for such design constraints can be graphically expressed in the drawing as shown in Fig. 12.a. Now suppose, for the sake of the example, that we are interested in modifying the house, and the origin of the right wall is selected and extended in the course of the interactive session, as shown in Fig. 12.b. In order to satisfy constraints (90) and (91) above, the position of the end extreme of wall_5 has to be translated to the intersection point of the two construction lines, as shown in Fig. 12.c.

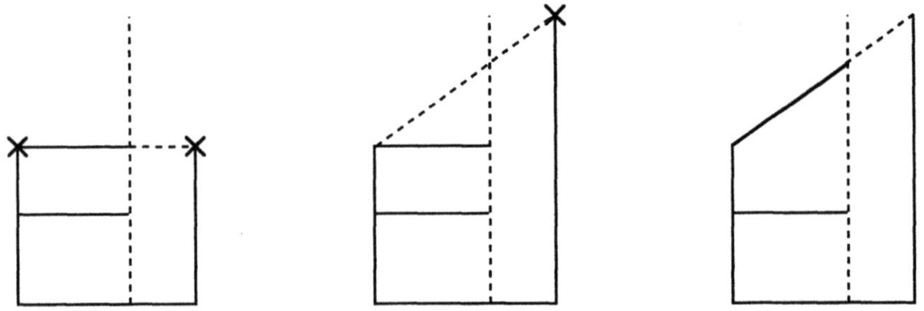

a) b) c)

Fig. 12. Modification of the house

Now we come to the definition of a transformation function that enforces constraints (90) and (91) in the course of an interactive session. Such a function can be defined as a Prolog clause that is proved by the interpreter at the end of every interactive cycle, before the interactive control is returned to the human-user. For the current example, the clause is as follows:

(92) **change_rule :- not collineal(wall_5, c_line_2),**
intersect_at(wall_5, c_line_1, POINT),
move_end(wall_5, POINT).

where collineal and intersect_at are the translations in Prolog of the corresponding geometrical predicate of **D** and move_end is a transformation function that changes the parametric position of the end-extreme of a line. The arguments of move_end are of types line and real_pair. This rule could be paraphrased as follows:

(93) **IF not collineal(wall_5, line_2) AND intersect_at(wall_5, line_1, POINT)**
THEN move_end(wall_5, POINT).

In the interpreter of GRAFLOG there are two clauses, namely move_end and move_origin, which modify the end and origin extremes of a line of the graphical representation. When a rule of this kind is identified by the interpreter a state transition takes places, and the change is reflected in the graphics too.

This example is an illustration of the kind of interactive manipulation that can be supported by an underlying graphical and logical representational structure $\mathbf{L_d}$. It is worth pointing out that different strategies for defining such a kind of constraints and transformation functions through the natural language and graphics dialogue can be envisaged. This facility of the design interpreter can be thought of as a very straightforward AI production system, but with a strong interface component. In the current prototype, human-users can define complex rule-based applications.

Furthermore, the rich semantics provided by the language $\mathbf{L_d}$ allows the definition of an additional layer of functionality. In our simple design domain, transformation functions are not only applied to drawings by the design interpreter: change rules can also be deduced by the system in the context of some design task. The method by which the system is able to induce such a kind of transformation rules has been developed under the heading of *intention propagation process* [8]. However, there is no room in this paper for presenting this process.

Conclusion

Here, we conclude this paper on the definition of the graphical and logical language for design $\mathbf{L_d}$. The language is useful for representing the semantics of drawings made out of dots, lines and polygons. The notions of extension and intension for drawings represented in $\mathbf{L_d}$ were discussed. We show how the representational system is integrated to the interactive interface and how the semantic interpretation of drawings is introduced by means of direct manipulation on drawings in the context of natural language expressions. The semantic representation of drawings is automatically deduced by the interpreter by means of a set of identification rules that act upon the graphical and linguistic input. We have also illustrated how the language can be used for the definition of causal relations between states in simple design tasks.

References

1. W.M. Beynon and A.G. Cohn, "Representing Design Knowledge in a Definitive Programming Framework," in *Pre-prints of the Second IFIP WG 5.2 Workshop on Intelligent CAD*, ed. H. Yoshikawa and T. Holden, The University of Tokyo, Tokyo, 1988.
2. D.R. Dowty, R.E. Wall, and S. Peters, *Introduction to Montague Semantics,* D. Reidel Publishing Company, Dordrecht, Holland, 1981.
3. G. Frege, "On Sense and Reference," in *Translations from the Philosophical Writings of Gottlob Frege*, ed. P.T. Geach and M. Black, p. 56, Blackwell, Oxford, 1952.
4. J.A. Goguen, J.W. Thatcher, and E.G. Wagner, "An Initial Algebra Approach to the Specification, Correctness, and Implementation of Abstract Data Types," in *Current Trends in Programming Methodology*, ed. R.T. Yeh, vol. IV, pp. 80-149, Prentice-Hall, 1978.
5. L.A. Pineda, *Un Algoritmo General para la Comparación de Polígonos,* Instituto Tecnológico de Monterrey, México, 1986. M.Sc. dissertation (in Spanish)

6. L.A. Pineda, E. Klein, and J. Lee, ''GRAFLOG: Understanding Graphics Through Natural Language,'' *Computer Graphics Forum*, vol. 7, no. 2, 1988.

7. L.A. Pineda, ''A. Compositional Semantic for Graphics,'' in *Eurographics'88 Conference Proceedings*, ed. D. Duce and P. Jancene, North-Holland, Amsterdam, 1988.

8. L.A. Pineda, *GRAFLOG: A Theory of Semantics for Graphics with applications to Human-Computer Interaction and CAD Systems*, 1989. Ph.D. thesis, University of Edinburgh

9. F.P. Preparata and M.I. Shamos, *Computational Geometry: an Introduction,* Springer-Verlag, Berlin, 1985.

10. R. Reiter, ''On Reasoning by Default,'' in *Readings in Knowledge Representation*, ed. R.J. Brachman & H.J. Levesque, pp. 402-410, Morgan Kaufman Publishers Inc., Los Altos, California, 1985.

11. M.I. Shamos, *Computational Geometry,* University Microfilms International, 1978. Ph.D. Thesis, Yale University

12. I.E. Sutherland, ''Sketchpad: A Man-Machine Graphical Communication System,'' in *AFIPS SJCC Proceedings*, pp. 329-346, 1963.

13. B. Tilove, ''Set Membership Classification: A Unified Approach to the Geometric Intersection Problem,'' *IEEE Transactions on Computers*, vol. C-29, no. 10, 1980.

14. K. Weiler, ''Polygon Comparation using Graph Representations,'' *ACM Siggraph'80 Conference Proceedings, Computer Graphics*, 1980.

9

An Experience with Developing a Design Knowledge Representation Language

Tetsuo Tomiyama, Deyi Xue, and Yoshiki Ishida

Department of Precision Machinery Engineering
Faculty of Engineering, The University of Tokyo
Hongo 7-3-1, Bunkyo-ku, Tokyo 113, Japan

Abstract: This paper describes our experience with developing a design knowledge representation language named IDDL (Integrated Data Description Language) on the object-oriented language Smalltalk-80.[†] IDDL is the kernel language of the IIICAD (Intelligent Integrated Interactive CAD) system and is developed to represent design knowledge about both design processes and design objects. First, the IDDL language specifications are reviewed. Following a brief note about its implementation, we then show a few examples illustrating how IDDL represents design knowledge.
Keywords: Intelligent CAD, Knowledge Engineering, Design Knowledge, Object Oriented Programming, Logic Programming, Modal Logic, Default Reasoning, ATMS, User Interface.

1. Introduction

Computer aided design (CAD) systems now play a crucial role in many engineering fields. Particularly in mechanical engineering, excellent tools to deal with geometrical information have been developed and designers are liberated from laborious drafting and can concentrate on more creative work. Recent advances in artificial intelligence and knowledge engineering technology presented a new direction for CAD, i.e., so-called intelligent CAD systems (ICAD). There are already enough systems to indicate that it is not too early to say that this trend is a definitely promising approach to releasing a designer's power to more essential and creative phases of design. This is perhaps the only way to achieve high productivity and quality in design at the same time.

Two different approaches can be observed in developing ICAD [1]. One is top-down and somehow abstract, and it insists that the system should be established on a sound theoretical basis to describe design knowledge; otherwise it contends, development of ICAD will be

[†] Smalltalk-80 is a registered trademark of Xerox Corporation.

overwhelmed by the complexity of design knowledge. In this approach, one might begin with analyzing the nature of design to arrive at a general platform or framework on which ICAD elements should be built. The other approach is bottom-up and rather concrete in that intelligent design systems for a narrow problem domain are first built and then connected to each other to form a whole system. This approach puts more emphasis on analyses of how design is performed using design knowledge and it often leads to development of so-called expert design systems or knowledge based design systems.

The IIICAD (Intelligent Integrated Interactive CAD) project was proposed at the Centre for Mathematics and Computer Science (CWI) in Amsterdam [13]. We observed that current CAD systems suffer from the following three problems:

- Not intelligent; e.g. the system does not understand the domain knowledge in the same context as the designer.
- Not integrated; e.g. the system does not support the entire design process, from the conceptual design to the production design.
- Poor in man-machine communication.

The core of the IIICAD project is the development of IDDL (Integrated Data Description Language), a design knowledge representation language that can describe both design processes and design objects [20]. Fig. 1 shows the configuration of the IIICAD system in which the *supervisor* controls design processes based on *scenarios* written in IDDL, and all the subsystems communicate with each other in IDDL. In this regard, IDDL is a framework to represent design knowledge and the IIICAD project takes, therefore, a top-down, framework approach.

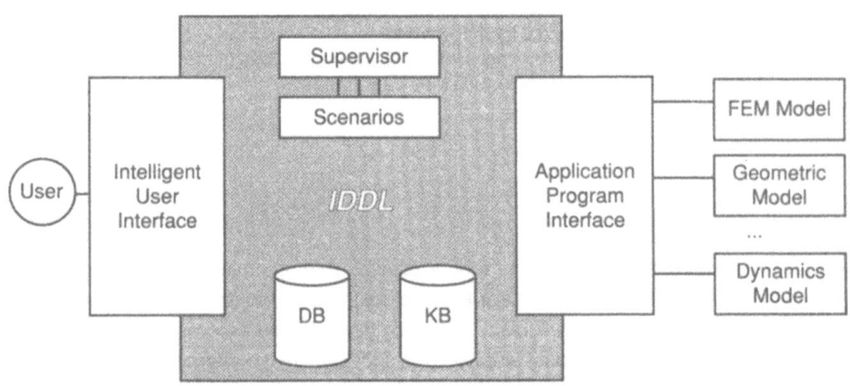

Fig. 1. The configuration of the IIICAD system

The original ideas of IDDL were first presented in [20] by the IIICAD project group at CWI. The present paper is a report about our experiences with IDDL) in an extended and modified version of the IIICAD project which is currently conducted at the University of Tokyo. The paper [19] in this volume describes another development of IDDL (ADDL, Artifact and Design Description Language) conducted at CWI. Although these two versions of IDDL (*viz.* IDDL and ADDL) are developed from philosophically the same ideas, there exist differences in minor details of both the specifications and implementations. (ADDL's specifications are found in [17].)

The remainder of the paper is organized as follows. In §2, we briefly review the IDDL specifications which are elaborated from a working draft in [20]. §3 describes techniques used for the implementation of IDDL. Most of the useful ideas for implementation of IDDL were taken from the work by Megens at CWI [8] which was, however, only partially finished. §4 is our contribution: we describe various types of design processes in IDDL with examples and evaluate the original ideas of IDDL from the design knowledge representation viewpoint. §5 concludes the paper.

2. IDDL Language Specifications

2.1. Overview

In the IIICAD project it is emphasized that such an advanced system should be constructed on a sound theoretical basis, and one of the project's goals is to establish a *theory of* CAD consisting of theories of design processes, design objects, and knowledge [20]. There are commercially available so-called ICAD systems, but most of these systems are much concerned with constraint information. These systems can deal with prototypical design objects; the structures of these objects are predefined in one way or another and their attributes construct a network of such relationships as equality, equivalence, derivation, and implication. This can be justified in a situation where the purpose of design is to modify details of once designed objects and the majority of design activities are to solve interrelated, complicated constraints, such as geometrical constraints in an intelligent manner. This is the case in mechanical engineering design that is one of applications of IIICAD. However, this point of view misses one of the most crucial aspects of design; i.e., design processes. Without explicit knowledge about design processes, for example, it might be extremely difficult to handle the trial-and-error nature of design, because such a process cannot be very effectively performed by simple backtracking.

There are three major concepts that are relevant to design and are represented in IDDL, viz. design objects (entities), their attributes, and relationships among entities. There is an assumption that a design process is a stepwise refinement process and is decomposed into subprocesses. In other words, a design process is composed of a set of subprocesses and how these subprocesses are combined determines the process.

IDDL takes a logical standpoint; i.e., design knowledge is represented by a set of facts and rules that are denoted by logical formulae and a design solution is obtained as the result of logical reasoning with these formulae. Thus, a design process must be (and can be) formalized logically [11, 18]. In our formalization, an entity is denoted by a term (both constant and variable) and a relationship between two entities is represented by a binary first-order predicate. Let us call this way of formalization an *extensional* description [14]. Any design knowledge about these predicates are formalized as simple rules to manipulate them. However, there can be cases where one may want to express and handle attributes of entities. In principle, such an attribute must be expressed by the equality predicate over the attribute and the value in an extension description method as follows. For instance, suppose the entity X is a box and has the attribute *depth* which is equal to 40 cm. The extensional description of this fact is given by

box(X) & depth(X, DepthOfBoxX) & =(DepthOfBoxX, 40).

In order to explicitly state the fact that the attribute belongs to the entity, a function (depth) is used. This is represented in IDDL as follows (*intensional description*):

box(X) & =(depth[X], 40).

This description method is convenient for computational and representational purposes, but it has a drawback as well: It assumes *structure*; i.e., that any box (whatever the box is) has depth, rather than height, for example. Otherwise, if used carelessly, any statement that assumes a box has the attribute *depth* becomes meaningless.

The following requirements were taken into consideration during the design of IDDL [20].

1. IDDL should be able to describe design knowledge both about design objects and design processes.
2. IDDL should be able to describe design objects, their relationships, and their attributes.
3. IDDL should be able to describe design processes, which are often performed in a trial-and-error manner and start with some assumptions.

These issues resulted into the following properties of IDDL. In a word, IDDL is a language which has flavors of both the object-oriented programming paradigm and the logic programming paradigm.

1. IDDL uses *scenarios* to describe design processes. A scenario is a meaningful chunk of design procedures, such as design of a particular mechanism, and decision making based on a certain technique.
2. *Objects* denote entities and *facts* are logical formulae that represent relationships among entities. *Worlds* represent a set of objects now being considered and, in fact, are partitions in the database.
3. IDDL has a multi-world mechanism based on *modal logic.*
4. There is a distinction between real facts and assumed facts. This feature allows *default reasoning* based on ATMS (Assumption-based Truth Maintenance) [5].

Fig. 2 illustrates how an IDDL *program* is executed. First, we describe design processes in terms of scenarios and functions. The results of its execution are stored in the database of the *active world* as facts, objects, and attributes of objects. The complete IDDL syntax is found in the Appendix.

2.2. Scenarios — descriptions of design processes

The construction of a scenario is shown in Fig. 3. By the object declarations and function declarations, objects or functions used in this scenario are declared. The main body of a scenario is the rule section. A rule has the well-known

IF <condition> THEN <operation>

structure in which, if the condition part is satisfied, then the operation part is executed. Execution of the operation part includes querying to the database, asserting facts into the database, and executing built-in predicates.

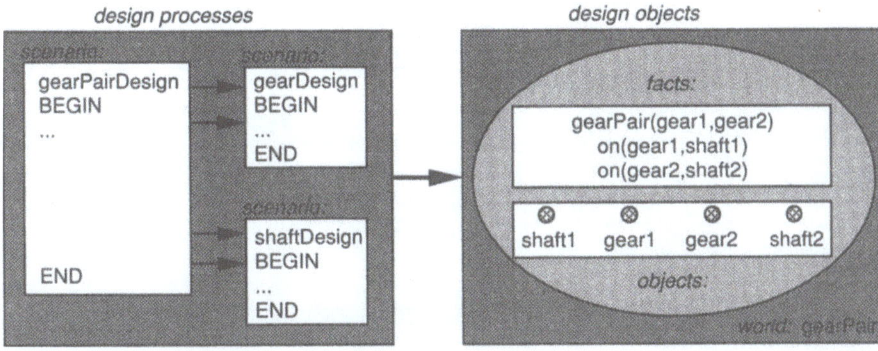

Fig. 2. Execution of an IDDL program

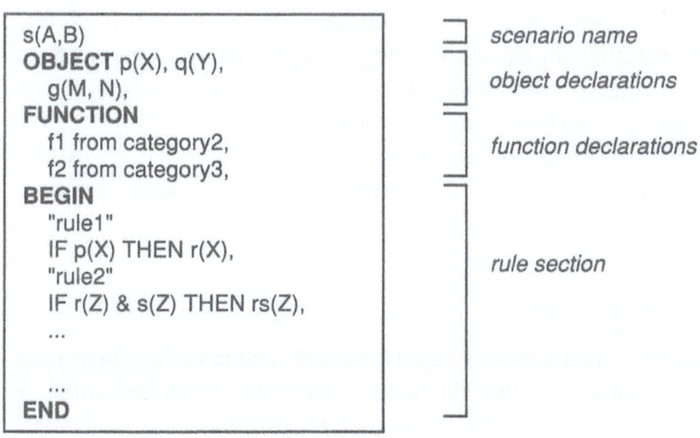

Fig. 3. A scenario

In a scenario, we can call other scenarios by using the built-in predicate select.[1] The execution of a scenario terminates in one of the following three conditions; the *success* condition, the *fail* condition, or the *no-more-rule* condition which is treated as a success. If the execution of a scenario is a success, all facts derived from this scenario are saved in the database; if failed, all facts derived from this scenario are removed from the database.

2.3. Objects and worlds — descriptions of design objects

IDDL has the concepts of objects, attributes of objects, and facts, and worlds to describe design objects (Fig. 4). A world consists of objects and facts which describe relationships among these objects. A world is in fact a partition of the database to preserve the results of design. An object in IDDL must be first declared in the object declaration in a scenario (Fig. 4) and then can be

[1] In ADDL, the use built-in predicate is used.

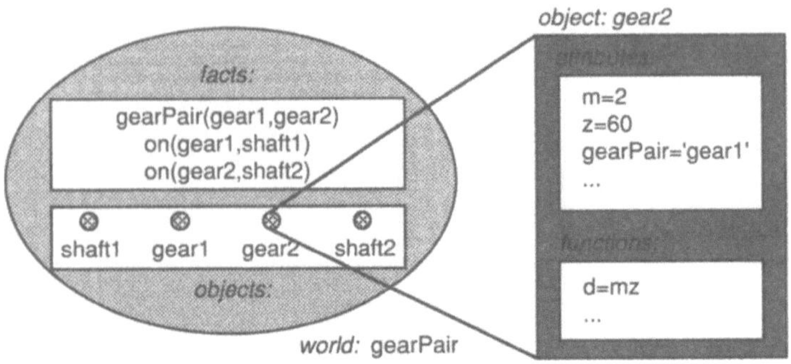

Fig. 4. An object

created at the time when an assertion of a fact that includes the object is made. An object must belong to a *type*. Since IDDL has no concepts of classes nor instances but *prototypical* objects based on *delegation* [7], there is instantiation in the sense of logic programming. Therefore, an object behaves much the same as an object in Smalltalk-80: it has attributes and we can define operations over these attributes (see the next section about functions), but by virtue of the delegation mechanism it can represent different types. This means that an object is created by copying the internal structure of the type. A name beginning with an upper case letter denotes a variable object, whereas one beginning with a lower case letter denotes a constant object.

A world can be created by the built-in predicate enclose:

```
enclose(world1, object1, object2, object3).
```

This creates a new world called world1 which contains object1, object2, and object3. There must always be one world active in the database, called the *active world,* in which execution of scenarios takes place and the results are preserved. If the modal logic feature (see §2.5) is used, there are accessible worlds (or worlds that can be seen) besides the current active world. However, even in such a case, there is always only one current active world. The current active world can be changed by using the built-in predicate enter or exit (Fig. 5).

The specialization mechanism of IDDL is based on *delegation* that allows much more useful specialization than widely-used *inheritance* [15]: i.e., the binding between objects is done imperatively by object declarations in the scenario, and an object belonging to a particular world can sometimes be bound to an object of a different type in a different world. (Contradictions created by this operation will not be automatically detected, unless relevant objects are sent to a certain world by using the built-in predicate sendObject (see §4.4.1) and the integrity of the database checked explicitly.)

Objects have internal data structure and are created by the built-in predicate makeObject; e.g.

```
makeObject(Gear, gear1)
```

where Gear is the type of the generated object gear1. As Fig. 6 illustrates, the internal data structure of an object is a collection of attributes and these attributes are accessed only by

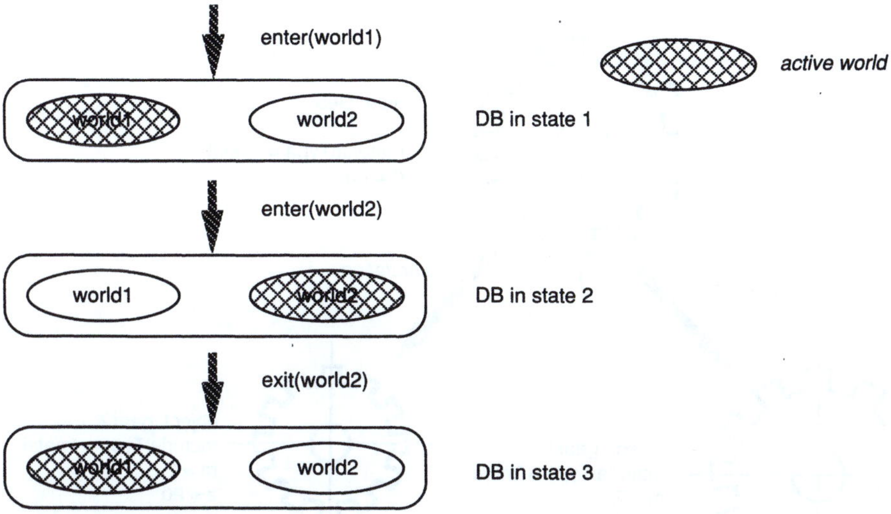

Fig. 5. Changing worlds

functions. This way of describing objects allows intensional descriptions (§2.1) and specialization which is one of the features of object-oriented languages. However, the type of an object is not a necessary condition to creat it, because there can be *general-type* objects that are almost typeless.

2.4. Functions — operations to design objects

Another important feature of the object-oriented paradigm is message passing [10] which is performed by functions in IDDL. The logic programming features of IDDL, such as the use of predicates and rules, are appropriate for representing static relationships among objects, whereas message passing can be used to describe dynamic behaviors of objects [15]. Functions of IDDL are used for the following purposes.

1. Function calling as message passing activates a procedural process of the receiver(s) that represents their dynamic behaviors.
2. As described in the previous section, objects have attributes and functions can be used to access these internal attributes. For instance, a point-type object can have its coordinates as attributes which can be accessed by functions.

Functions are defined outside the scenario and imported by the function declaration of a scenario into the scenario.

1. Functions defined as attributes of objects. For example,

   =(depth[gear1], 40),

   says that the value of the depth function of gear1 is equal to 40. In other words, objects have internal memory spaces for those attributes.

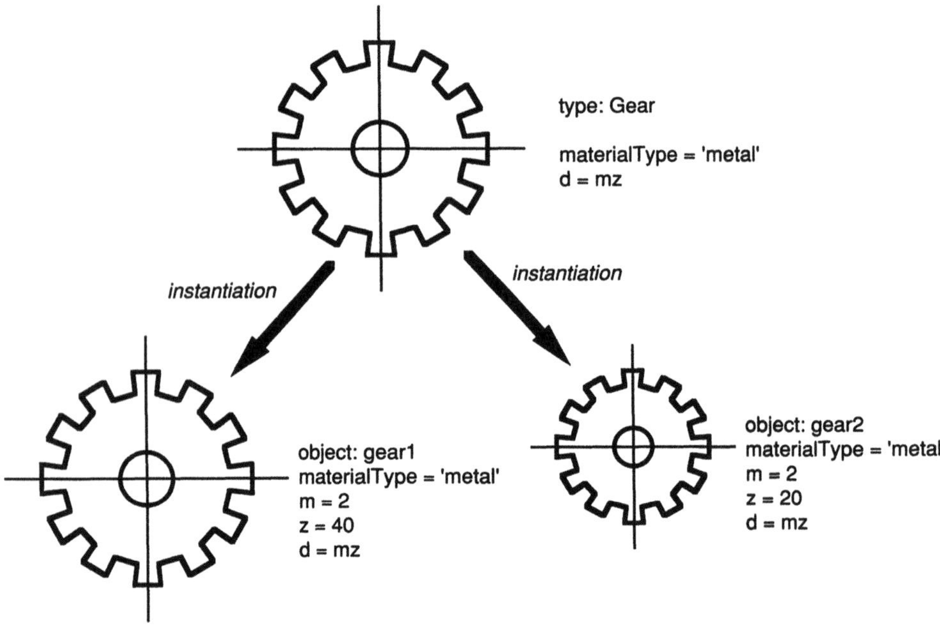

Fig. 6. Object Types

2. Functions defined commonly over a set of objects. For instance, a function to compute the distance between two points, A and B, is given by:

```
distance[A, B]
    X := x[A] – x[B],
    Y := y[A] – y[B],
    D := sqrt[sqr[X] + sqr[Y]],
↑D
```

   where x, y, sqrt, and sqr are functions and ↑ denotes "return the following value." Although we can use temporary variables, such as X, Y, and D of this example, their values are not stored. (If the value must be stored, we must use attributes of objects.)

2.5. Modal logic

Consider a design problem such as "If one of the following conditions is satisfied, the entire design should be considered as a success." IDDL supports it by introducing modal logic with necessity and possibility modalities.

IDDL has two modal operators, viz. #N for necessity and #P for possibility.[2] For example,

---

[2] ADDL has only the P modal operator which works in exactly the same way as the default operator #D of IDDL. In ADDL, the modal logic capability of IDDL is provided by the multiple world mechanism [17, 18].

consider #Np(a). If all worlds *accessible* from the current world have the fact p(a), #Np(a) is true; otherwise, it is false. Similarly, matching #Pp(a) to the database returns true if at least one of the accessible worlds has the fact p(a); otherwise, it returns false.

In order to determine which world is *accessible* (i.e., *can be seen*), we need to choose the most appropriate kind of modal logic systems from T, B, S4 or S5 systems according to the accessibility properties including reflexivity, symmetry, and transitivity (see e.g. [4]). The definitions of accessibility properties and modal logic systems are shown in Tables 1 and 2. For example, the modal logic system B has the property of symmetry. In such a case, if it is defined that the world w_1 is also accessible from the world w_2, the system will automatically conclude that w_2 is accessible from w_1. These properties are defined by the built-in predicates multiWorldsSystem and canSeeWorlds to choose the most appropriate modal logic system and accessibilities among worlds.

reflection	A world can see itself
symmetry	If world A can see world B, then world B can see world A as well
transitivity	If world A can see world B and world B can see world C, then world A can see world C as well

Table 1. The accessibility properties

system name	properties
T	reflection
B	reflection, symmetry
S4	reflection, transitivity
S5	reflection, symmetry, transitivity

Table 2. Modal logic systems

2.6. ATMS

IDDL employs the default operator #D to allow database management based on ATMS [5]. IDDL distinguishes real facts and assumed facts. An assumed fact has descriptions showing from which facts it was derived. When a real fact is added to the database, assumed facts are checked for the integrity of the database; assumed facts may be changed to real facts or removed from the database, depending whether they agree or conflict with newly added real facts (see §4.4.3).

3. Implementation of IDDL

The IDDL language is implemented in Smalltalk-80 [3]. Smalltalk-80 was employed as the implementation language for the following reasons:

1. IDDL itself is a combination of the logic programming and the object-oriented programming paradigms [15, 20]. Most of the properties of IDDL, such as delegation, can be easily implemented using the features of Smalltalk-80.
2. Smalltalk-80 provides facilities to easily build an excellent user interface environment.
3. Smalltalk-80 itself is a good tool for rapid prototyping that allows incremental programming.

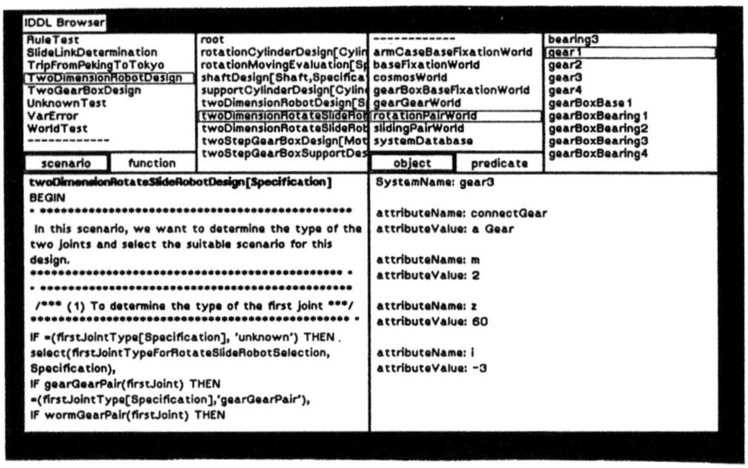

Fig. 7. IDDL **browser**

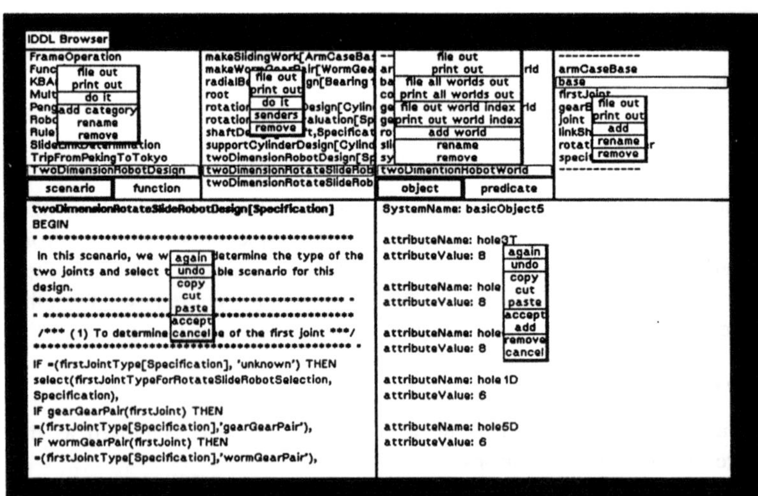

Fig. 8. Menus of the IDDL **browser**

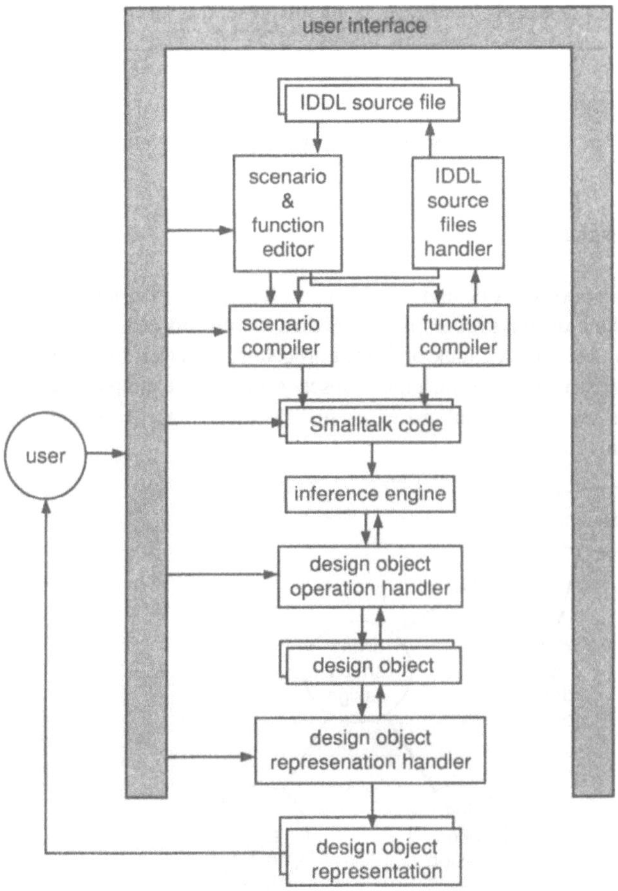

Fig. 9. The architecture of the IDDL system

The programming environment of IDDL is the IDDL *browser* shown in Fig. 7. Fig. 8 shows available menus on the IDDL browser. On the left side of the browser window, we can write and compile scenarios and functions and eventually execute scenarios. On the right side of the browser window, we can see the results of the execution described in terms of worlds, objects, attributes of objects, and facts.

Fig. 9 depicts the architecture of the IDDL system. There are two compilers, viz. *scenario compiler* and *function compiler,* to translate scenarios and functions into Smalltalk-80 code, respectively. Therefore, an IDDL object is a Smalltalk-80 object. This idea was taken from the work by Megens [8]. The *scenario & function editor* is used for editing IDDL source code. The *inference engine* sends messages to the *design object operation handler* to operate objects and the database. The *design object representation handler* is used for presenting the status of the database.

4. Representing Design Knowledge in IDDL

The following sections are based on an analysis of design knowledge in an intelligent CAD environment presented in [12].

4.1. Representing design processes in IDDL

4.1.1. Design processes and subprocesses. A design process can often be divided into several subprocesses. The result of the whole design process is obtained as a collection of solutions of the subprocesses. Since a scenario describes a chunk of design procedures, this nature of design can be represented by the scenario structure calling other subscenarios. For example, we can divide the design of diving clothes into several subprocesses such as the design of the hat, the coat, the trousers, and the shoes (Fig. 10).

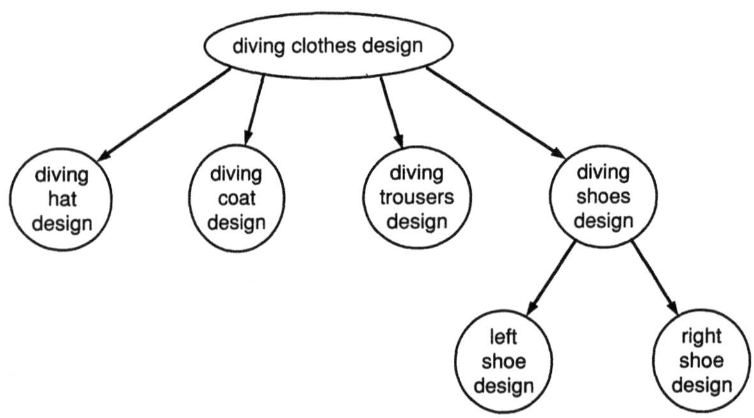

Fig. 10. Design subprocesses

4.1.2. Design knowledge and meta design knowledge. Design is not a simple problem solving process; rather, it also includes a process to determine the way to solve the problem, to find out which knowledge should be used, etc. This means that we need meta knowledge (or knowledge about knowledge) to control a design process. IDDL allows to write meta scenarios to select appropriate lower-level scenarios.

4.1.3. The trial-and-error method. To solve a problem, we often use the trial-and-error method. Using the multi-world mechanism of IDDL (§2.5) and its default reasoning feature, we can develop alternative solutions in different worlds until we arrive at satisfactory solutions.

4.1.4. Application of modal logic. IDDL allows a designer to develop alternative different worlds at the same time. We use the #N operator to describe conditions common to all related worlds and the #P operator to describe conditions which hold in at least one of the worlds. In the example of Fig. 10, we can design components of the diving clothes separately in different worlds and the waterproof condition can be added by saying #Nwaterproof.

4.2. Representing design objects in IDDL

4.2.1. Design objects from multiple viewpoints. The multi-world mechanism allows to describe the same object from different points of view. An object can have different user names (Fig. 11 (a)) and different properties including type in different worlds. (However, as mentioned before, semantical conflicts caused by having different types for the same object are not automatically detected.) This allows viewing an object from multiple viewpoints, which is essentially useful for mechanical engineering design. On the other hand, it is also possible to have different objects under the same name in different worlds (Fig. 11 (b)).

4.2.2. Database integrity. The IDDL system incorporates an ATMS to keep the integrity of its database. As a design proceeds, it sometimes happens that initial assumptions must be changed. In such a case, the facts derived from these assumptions marked with the default operator #D must also change. When a fact is added into the database, all related facts will be automatically checked in order to maintain the consistency of the database (see §2.6). Note that this check is done with respect to integrity within a single world and that it is different from the conflict check concerning the multi-world mechanism.

4.3. Intelligent design interface

IDDL is merely a language for system designers, so to speak, and not for domain design experts. To be used by designers directly, we provide an intelligent design interface on top of IDDL in the IIICAD architecture. The primary step we take is to make IDDL *programming* interactive. Fig. 12 shows a system called *design browser* for this purpose. This browser has the following abilities:

1. Drawing figures and inputting data, such as attributes, types, and names of objects identified by figures.
2. Selecting adequate scenarios to be executed.
3. Displaying output, such as figures and results of IDDL programs.
4. Storing information about the designed object, such as design models, the initial input, and the order of execution of the scenarios.

A future, advanced step might be to develop a graphical interface which extracts a designer's concepts from interactions to create graphical objects and which translates them into both design process knowledge as IDDL scenarios and design object knowledge as IDDL objects. This approach is expected to contribute to solving the knowledge acquisition problem.

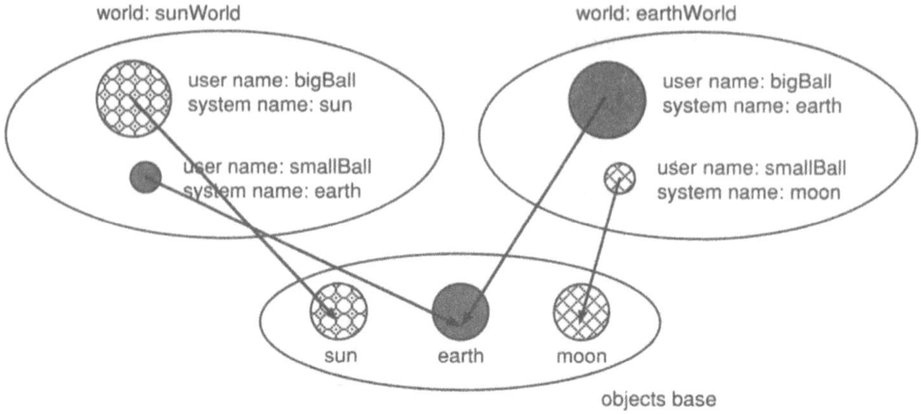

(a) Same objects with different user names in different worlds

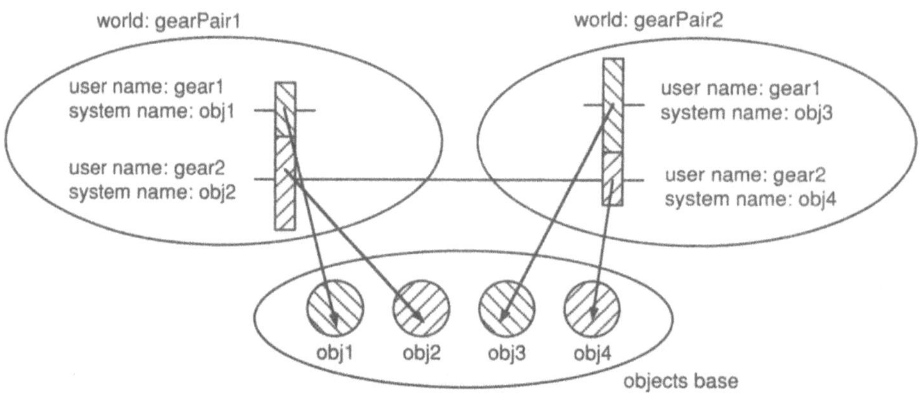

(b) Different objects with the same user names in different worlds

Fig. 11. User names and system names of objects

4.4. Examples

4.4.1. Representing design objects by using worlds and objects. A simple example of mechanical design, the design of a gear box, is illustrated here. We have a scenario for a gear box and a subscenario for a gear pair. Because the same scenario was executed twice, two objects with the same user name are first generated. (Of course, since the system distinguishes user names and system names, this is not a problem.) Second, gear1 is sent from gearPairWorld1 to linkWorld with the user name gearA using the built-in predicate sendObject. In the same way, we create gearB and fix gearA and gearB on the same shaft. Third, we determine the parameters of the four gears (i.e. attributes) and complete this design (Fig. 13 (a)). The designed gear box has not only attributes but also the internal relationships that represent its structure.

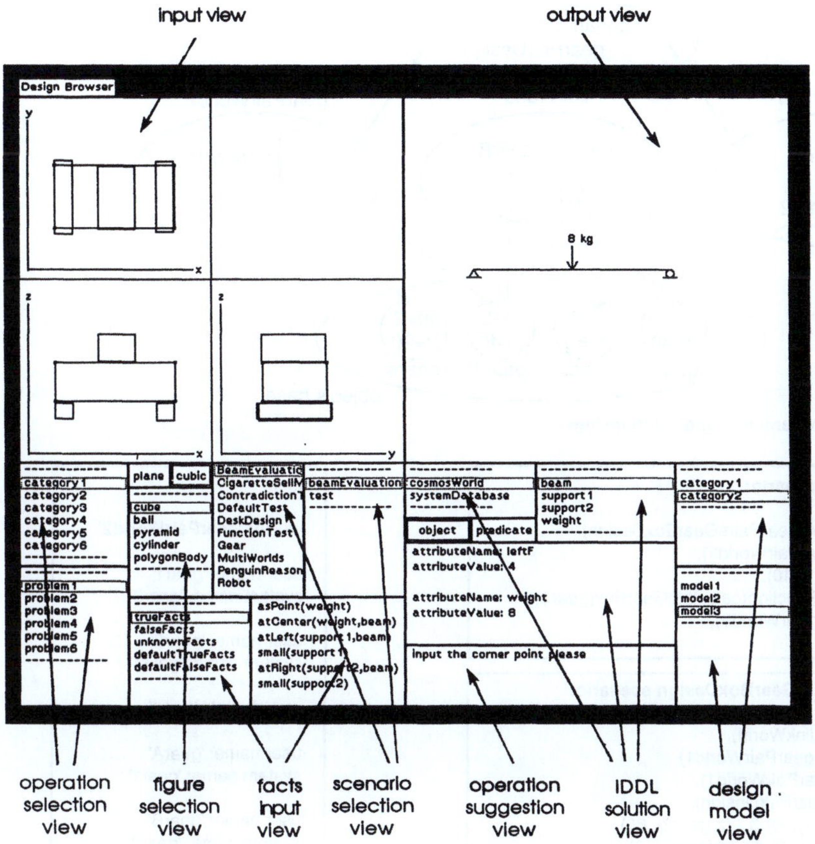

Fig. 12. Design browser

4.4.2. Representing designed objects in different models. Fig. 14 depicts an example for evaluation of a beam. First, as shown in Fig. 14 (b) to (d), the boundary conditions are determined. The rules used in this example are shown in Fig. 14 (e) and the results are obtained as shown in Fig. 14 (f) and (g). This example shows most clearly how a design process proceeds on the design browser which is an application of the IDDL browser. A designer is always allowed to examine the database to check the results. If something is wrong, she/he can interactively change whatever is needed including IDDL code. However, this might not be the best, because the current version of IDDL does not really provide a user-friendly interface.

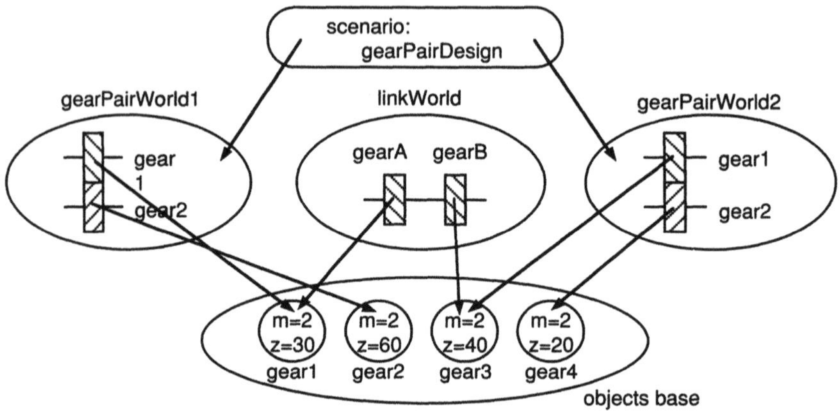

(a) Determining parameters and relationships

```
example3 scenario:
  BEGIN
    select(twoGearPairsGearBoxDesign),
    enter(gearPairWorld2),
    =(n(gear1),40),
    proceedFunction(calculateGearBox(gear1)),
    exit(gearPairWorld2),
  END
```

```
twoGearPairsGearBoxDesign scenario:
  BEGIN
    enclose(linkWorld),
    enclose(gearPairWorld1)
    enter(gearPairWorld1),
    select(gearPairDesign),
    =(z(gear1),30) & =(z(gear2), 60),
    =(m(gear1),2) & =(m(gear2), 2),
    sendObject(gear1,linkWorld,gearA),
    exit(gearPairWorld1),
    ...
    ...
    enter(linkWorld),
    proceedFunction(linkGear(gearA, gearB)),
    exit(linkWorld),
  END
```

```
gearPairDesign scenario:
  BEGIN
    makeObject(Gear, gear1),
    makeObject(Gear, gear2),
    proceedFunction(gearPair(gear1, gear2)),
  END
```

(b) IDDL program

```
... ...

world: 'gearPairWorld2'

user name: 'gear1'
system name: 'gear3'

user name: 'gear2'
system name: 'gear4'

world: 'linkWorld'

user name: 'gearA'
system name: 'gear1'

user name: 'gearB'
system name: 'gear3'
```

```
system object name: 'gear1'

attribute name: m
attribute value: 2

attribute name: linkGear
attribute value: gear3

attiribute name: n
attribute value: 40

... ...
```

(c) Result of the execution

Fig. 13 A gear box design

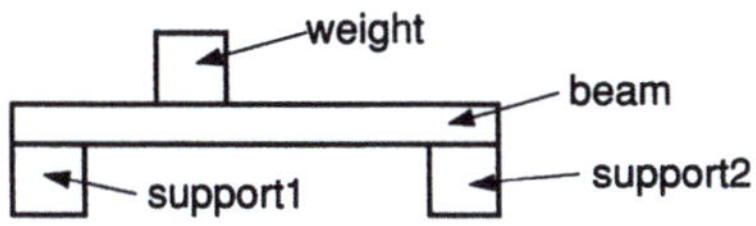

(a) The beam structure

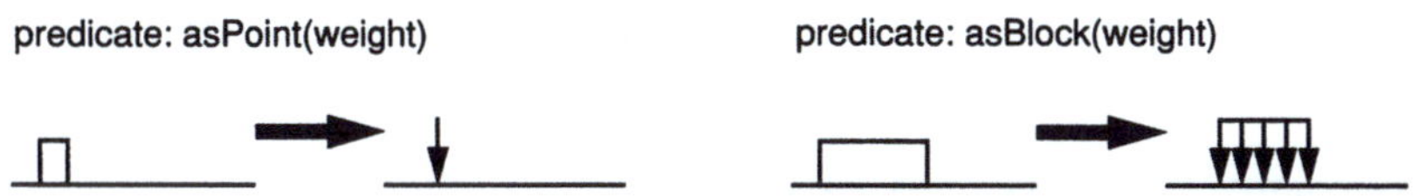

(b) Determination of the type of load

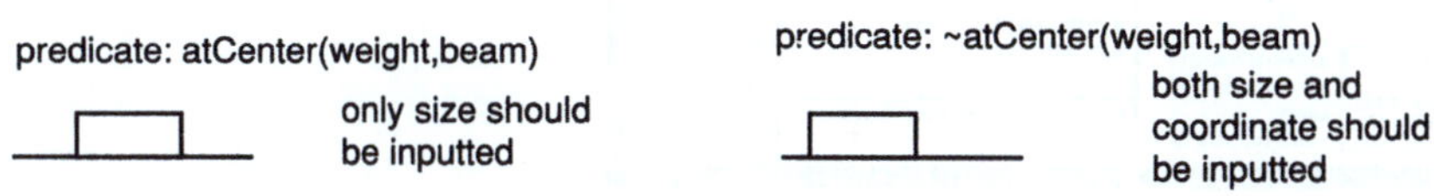

(c) Determination of the location of the load

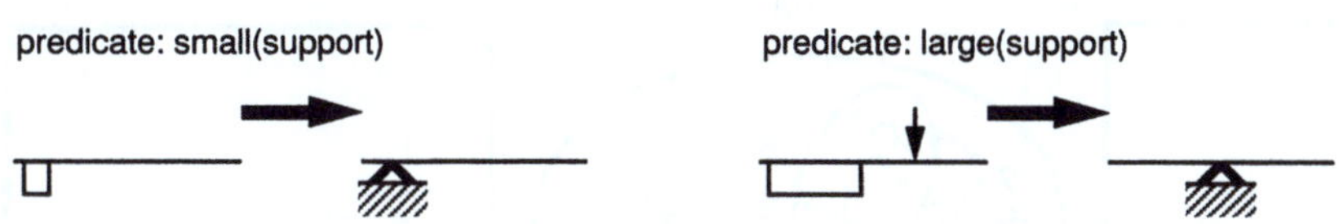

(d) Determination of the support condition

```
BeamEvaluation scenario:
  BEGIN
    IF asPoint(weight) & atCenter(weight, beam) THEN proceedFunction(carryWeight1(beam, weight)),
    IF asPoint(weight) & ~atCenter(weight, beam) THEN proceedFunction(carryWeight2(beam, weight)),
    IF asBlock(weight) & atCenter(weight, beam) THEN proceedFunction(carryWeight3(beam, weight)),
    IF asBlock(weight) & ~atCenter(weight, beam) THEN proceedFunction(carryWeight4(beam, weight)),
    IF atLeft(support1, beam) & small(support1) THEN proceedFunction(smallLeftSupport(beam, support1)),
    IF atLeft(support1, beam) & large(support1) THEN proceedFunction(largeLeftSupport(beam, support1)),
    IF atRight(support2, beam) & small(support2) THEN proceedFunction(smallRightSupport(beam, support2)),
    IF atRight(support2, beam) & large(support2) THEN proceedFunction(largeRightSupport(beam, support2)),
... ...
```

(e) Rules

Fig. 14. Evaluations of a beam

4.4.3. Data integrity inspection. Suppose we have the following rules:

```
rule1  IF T THEN #D p1(a) & #D q(a),
rule2  IF p1(X) THEN p2(X),
rule3  IF p2(X) THEN p3(X),
rule4  IF p3(X) & q(X) THEN r(X),
rule5  IF TTHEN p2(a),
rule6  IF T THEN q(a)
```

where T is a logical constant which is always true. Fig. 15 shows the data dependencies among

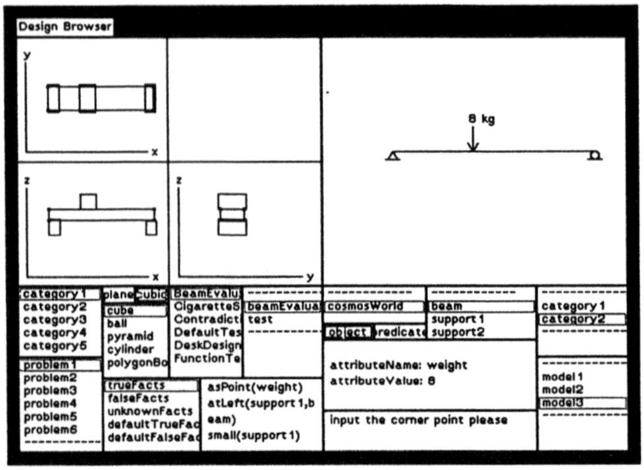

(f) A solution

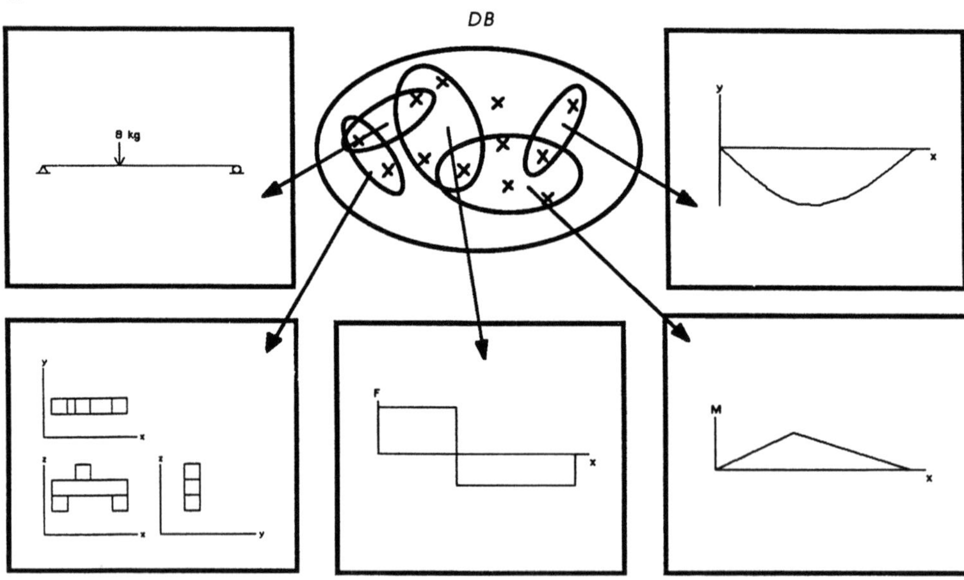

(g) Other solutions

Fig. 14. Evaluations of a beam (continued)

default facts and real facts. By executing rules rule1 to rule4, we get five default facts in the database (Fig. 15 (a)). When rule5 was executed, the real fact p2(a) is added to the database. Because p2(a), p3(a), and r(a) are related with p2(a), these three default facts are checked and p2(a) and p3(a) will be changed into real facts (Fig. 15 (b)). Similarly, when rule6 is executed, q(a) and r(a) will be removed from the database (Fig. 15 (c)).

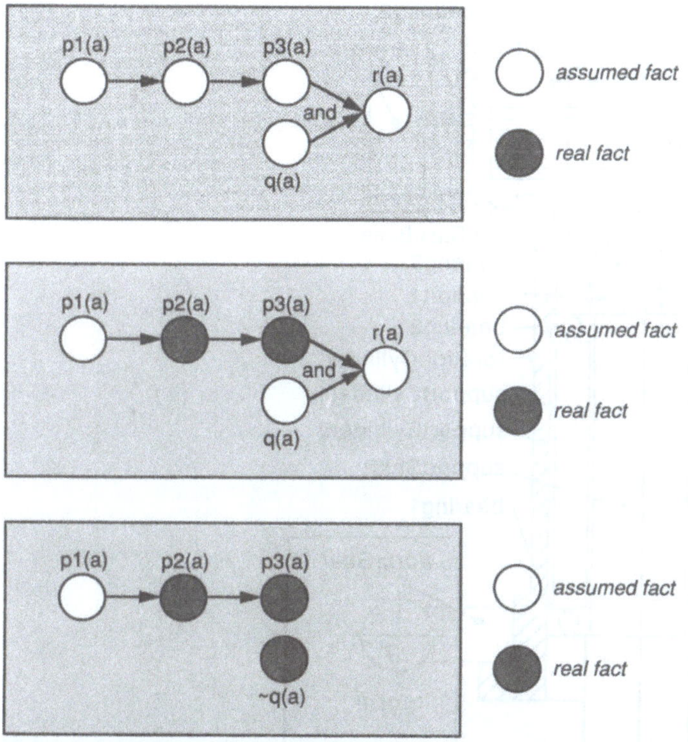

Fig. 15. Data dependencies in the database

4.4.4. Design of a two degrees-of-freedom robot. Fig. 16 illustrates a relatively complex design example of two degrees-of-freedom robots. The design process is described by the execution sequence (i.e. the sequence of the executed select built-in predicates) of the IDDL program in Fig. 17.

This example shows that IDDL has potential for describing a fairly wide range of design, provided the design process can be explicitly *downloaded* in IDDL. This essentially means that IDDL programming using the IDDL browser (Fig. 7) is equivalent to designing a new type of design objects, whereas just executing existing IDDL codes is a routine design.

5. Conclusions

This paper reports the development of an experimental version of IDDL (Integrated Data Description Language) for the IIICAD system implemented on Smalltalk-80. IDDL is meant to have the capability of describing both design processes and design objects. This, in our view, is the most characteristic missing in current ICAD research.

IDDL has the concepts of scenarios to represent design processes, objects for design objects, and functions for operations over objects. The concept of worlds is useful to represent an area of

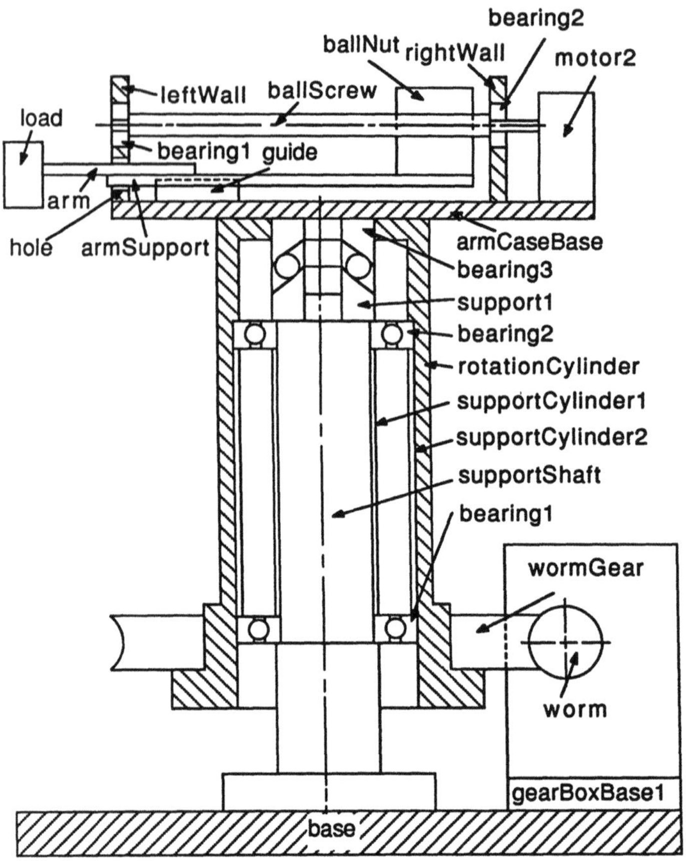

Fig. 16. The structure of the designed robot

interest or a situation in which a design currently takes place. The scenario mechanism together with the multi-world mechanism and the ATMS-like default reasoning facility allows various kinds of design techniques such as the trial-and-error method.

For future work, the following issues will be considered.

1. Research on more domain dependent design knowledge is needed, because IDDL provides only a framework to represent design knowledge. It must be clarified how real design processes are driven and by what kind of knowledge, e.g., in mechanical engineering. For this purpose, qualitative physics as a modeling framework may be necessary [2, 6, 9].
2. IDDL assumes only that design is a stepwise refinement process. This is not perhaps sufficient to model real designer's thought process and we need a more elaborated design process model, preferably, based on logic [11]. This may further necessitate a sophisticated design information management system beyond ATMS.

```
<1>: root
<2>: select(twoDimensionRobotDesign,specification)
<3>: select(twoDimensionRotationSlidingRobotDesign,Specification)
<4>: select(firstJointTypeForRotationSlidingRobotSelection,Specification)
<5>: select(rotationMovingEvaluation,Specification)
<6>: select(evaluateGearGearPair,Specification)
<7>: select(evaluateWormGearPair,Specification)
<8>: select(twoDimensionRotationSlidingRobotType1Design,Specification)
<9>: select(makeBaseFixationWork,base,supportShaft,Specification)
<10>: select(makeFixationWork,Shaft,Base)
<11>: select(determineThinPlateHoles,Object1,Object2,screw1,screw2,screw3,screw4)
<12>: select(determineThickPlateHoles,Object1,Object2,screw1,screw2,screw3,screw4)
<13>: select(makeGearBoxFixationWork,gearBoxBase1,base,Specification)
<14>: select(makeFixationWork2,GearBoxBase,Base)
<15>: select(determineThinPlateHoles,Object1,Object2,screw1,screw2,screw3,screw4)
<16>: select(determineThickPlateHoles2,Object1,Object2,screw1,screw2,screw3,screw4)
<17>: select(makeRotationWork,supportShaft,rotationCylinder,gearBoxBase1,Specification)
<18>: select(makeRotationPair,SupportShaft,RotationCylinder,Specification)
<19>: select(shaftDesign,Shaft,Specification)
<20>: select(radialBearingDesign,bearing1,bearing2,Shaft)
<21>: select(rotationCylinderDesign,Cylinder,bearing1,bearing2,Shaft)
<22>: select(axialBearingDesign,bearing3,Shaft,Cylinder)
<23>: select(supportCylinderDesign,supportCylinder1,supportCylinder2,Shaft,Cylinder,bearing1,bearing2)
<24>: select(fixWormGear,wormGear,RotationCylinder)
<25>: select(makeWormGearPair,wormGear,worm,Specification)
<26>: select(gearBox1IDetermination,Specification)
<27>: select(wormDesign,Worm,WormGear,Specification)
<28>: select(wormSupportBearingDesign,wormSupportBearing1,Worm)
<29>: select(wormSupportBearingDesign,wormSupportBearing2,Worm)
<30>: select(makeGearBox,motor1,Specification,GearBoxBase)
<31>: select(twoStepGearBoxDesign,Motor,Specification,GearBoxBase)
<32>: select(twoStepGearBoxSupportDesign,gear1,gear2,gear3,gear4,GearBoxBase)
<33>: select(makeFixationWork,rotationCylinder,armCaseBase)
<34>: select(determineThinPlateHoles,Object1,Object2,screw1,screw2,screw3,screw4)
<35>: select(determineThickPlateHoles,Object1,Object2,screw1,screw2,screw3,screw4)
<36>: select(makeSlidingWork,armCaseBase,Specification)
<37>: select(ballScrewSlidingPairDesign,ArmCaseBase,Specification)
<38>: select(caseBaseDesign,CaseBase,Specification)
<39>: select(twoWallsFixation,leftWall,rightWall,CaseBase,Specification)
<40>: select(makeBallScrewPair,ballScrew,ballNut,Specification)
<41>: select(armDesign,arm,armSupport,guide,rightWall,CaseBase,Specification)
<42>: select(guideDesign,Guide,CaseBase,RightWall)
<43>: select(holeDesign,hole,leftWall,arm,armSupport,guide)
<44>: select(ballScrewSupportDesign,ballScrew,leftWall,rightWall)
<45>: select(gearBox2Design,CaseBase,Specification)
```

Fig. 17. The execution sequence of the IDDL program

3. Although one of the main goals of IIICAD is the integration of design knowledge [13], the current prototype of IDDL does not have such an ability. For this, we need to incorporate the metamodel mechanism which integrates various kinds of models of design objects in a single, uniform framework [12, 16].

Acknowledgements

We would like to thank the students in the Yoshikawa-Tomiyama Laboratory, Department of Precision Machinery Engineering, Faculty of Engineering, The University of Tokyo, for their help and constructive comments on this study. Especially, we would like to mention that Susumu Hamada made important contributions by compiling the IDDL language specifications from previously unorganized material and by providing good working examples. We would like to thank Professor Hiroyuki Yoshikawa of the University of Tokyo for his valuable advice and continuous support. We are grateful also to the reviewers and editors whose comments and suggestions were very helpful in revising the paper.

References

1. *Intelligent CAD Systems I: Theoretical and Methodological Aspects,* Springer-Verlag, Berlin , 1987.
2. B. Faltings, ''Qualitative Kinematics in Mechanisms,'' *Proceedings of IJCAI-87*, pp. 436-442, 1987.
3. A. Goldberg and D. Robson, *Smalltalk-80: The Language and its Implementation,* Addison-Wesley, Reading, 1983.
4. G.E. Hughes and M.J. Cresswell, *An Introduction to Modal Logic,* Methuen, London, 1968.
5. J. de Kleer, ''An Assumption-based TMS,'' *Artificial Intelligence*, no. 28, pp. 127-162, 1986.
6. K. Kurumatani, T. Tomiyama, and H. Yoshikawa, ''Qualitative Representation of Machine Behaviors for Intelligent CAD Systems,'' *Special Issue on Computational Theories of Design — Applications to the Design of Machines*, 1989.
7. H. Lieberman, ''Using Prototypical Objects to Implement Shared Behavior in Object Oriented Systems,'' *Proceedings of the ACM Conference on Object-Oriented Programming Systems Languages and Applications*, pp. 214-223, 1986.
8. M. Megens, An Implementation of a Simple Design Description Language, Master's thesis, Computer Science Department, University of Amsterdam, 1987.
9. S.S. Murthy and S. Addanki, ''PROMPT: An Innovative Design Tool,'' in *Expert Systems in Computer-Aided Design*, ed. J.S. Gero, pp. 323-347, North-Holland, Amsterdam, 1987.
10. M. Stefik and D.G. Bobrow, ''Object Oriented Programming: Themes and Variations,'' *AI Magazine*, vol. 6, no. 4, pp. 40-62, 1986.
11. H. Takeda, T. Tomiyama, and H. Yoshikawa, ''Logical Formalization of Design Processes for Intelligent CAD Systems,'' in *Intelligent CAD 2: Proceedings of the Second IFIP Working Group 5.2 Workshop on Intelligent CAD*, ed. H. Yoshikawa, and T. Holden, North-Holland, Amsterdam, in preparation, 1989.
12. T. Tomiyama and P.J.W. ten Hagen, ''Organization of Design Knowledge in an Intelligent CAD Environment,'' in *Expert Systems in Computer-Aided Design*, ed. J.S. Gero, pp. 119-147, North-Holland, Amsterdam, 1987.
13. T. Tomiyama and P.J.W. ten Hagen, ''The Concept of Intelligent Integrated Interactive CAD Systems,'' CWI Report No. CS-R8717, Centre for Mathematics and Computer Science, Amsterdam, 1987.

14. T. Tomiyama and P.J.W. ten Hagen, "Representing Knowledge in Two Distinct Descriptions: Extensional vs. Intensional," CWI Report No. CS-R8718, Centre for Mathematics and Computer Science, Amsterdam, 1987.

15. T. Tomiyama, "Object Oriented Programming Paradigm for Intelligent CAD Systems," in *Intelligent CAD Systems II: Implementational Issues*, ed. V. Akman, P.J.W. ten Hagen, and P.J. Veerkamp, pp. 3-16, Springer-Verlag, Berlin, 1989.

16. T. Tomiyama, T. Kiriyama, H. Takeda, D. Xue, and H. Yoshikawa, "Metamodel: An Key to Intelligent CAD Systems," *Research in Engineering Design*, vol. 1, no. 1, pp. 19-34, 1989.

17. P. Veerkamp, V. Akman, P. Bernus, and P.J.W. ten Hagen, "IDDL: A Language for Intelligent Interactive Integrated CAD Systems," in *Intelligent CAD Systems II: Implementational Issues*, ed. V. Akman, P.J.W. ten Hagen, and P.J. Veerkamp, pp. 58-74, Springer-Verlag, Berlin, 1989.

18. P. Veerkamp, "Multiple Worlds in an Intelligent CAD Systems," in *Intelligent CAD 1: Proceedings of the First IFIP Working Group 5.2 Workshop on Intelligent CAD*, ed. H. Yoshikawa, and D.C. Gossard, North-Holland, Amsterdam, 1989.

19. P. Veerkamp, R. Pieters Kwiers, and P.J.W. ten Hagen, "Design Process Representation in IDDL," in *in this volume*, 1990.

20. B. Veth, "An Integrated Data Description Language for Coding Design Knowledge," in *Intelligent CAD Systems I: Theoretical and Methodological Aspects*, ed. P.J.W. ten Hagen, and T. Tomiyama, pp. 295-313, Springer-Verlag, Berlin, 1987.

APPENDIX: THE SYNTAX OF IDDL

```
<scenario>        ::= <declarations> <rule block> | <rule block>
<declarations>    ::= <object decl> <function decl> | <object decl>
                    | <function decl>
<object decl>     ::= OBJECT <object list>
<object list>     ::= <predicate> ,| <predicate> , <object list>
<function decl>   ::= FUNCTION <function list>
<function list>   ::= <external func> ,
                    | <external func> , <function list>
<external func>   ::= <func name> from <category name>
<category name>   ::= <identifier>
<rule block>      ::= BEGIN <rules> END
<rules>           ::= <rule> ,| <rule> , <rules>
<rule>            ::= <wff> |  <if> <wff> THEN <wff>
<if>              ::= IF | IFF | IFU
<wff>             ::= <fact> | ( <wff> ) | <wff> <binary l-op> <wff>
<fact>            ::= <general fact> | <modal l-op> <general fact>
<general fact>    ::= <basic fact> | <default l-op>  <basic fact>
<basic fact>      ::= <l-const> | <predicate>
                    | <unitary l-op> <predicate>
<predicate>       ::= <identifier> | <identifier> ( <term list> )
<term list>       ::= <expression> | <expression> , <term list>
<modal l-op>      ::= #N | #P
<default l-op>    ::= #D
<unitary l-op>    ::= ~ | %
<binary l-op>     ::= & | |
<l-const>         ::= T | F | U
<function>        ::= <func name> [ <term list> ] <function body>
<func name>       ::= <identifier>
<function body>   ::=  <local assign> <return value>
<local assign>    ::= <assignment> | <assignment> <local assign>
<assignment>      ::= <variable> :=  <expression> ,
<return value>    ::= ^ <expression>
<expression>      ::= <variable> | <number const> | <identifier>
                    | <char const> | <identifier> <expression>
                    | <function name> [ <term list> ] | ( <expression>)
                    | <expression> <identifier> <expression>
<identifier>      ::= <constant> | <comparing op> | <computing op>
<comparing op>    ::= = | < | > | <= | >= | <>
<computing op>    ::= + | - | * | /
<constant>        ::= <little letter> | <little letter> <char set>
<variable>        ::= <large letter> | <large letter> <char set>
<char set>        :: ASCII character set
<large letter>    ::= A|B|C|D|E|F|G|H|I|J|K|L|M|N|O|P|Q|R|S|T
                    |U|V|W|X|Y|Z
<little letter>   ::= a|b|c|d|e|f|g|h|i|j|k|l|m|n|o|p|q|r|s|t
                    |u|v|w|x|y|z
<digit>           ::= 0|1|2|3|4|5|6|7|8|9
<char const>      ::= ' <char set> '
<number const>    ::= <integer> | <real>
<number>          ::= <digit> | <digit> <number>
<integer>         ::= [-] <number>
<real>            ::= <integer>.<number> [e[+|-] <number>]
```

10

Design Process Representation in ADDL

Paul Veerkamp, Ravic Pieters Kwiers, and Paul ten Hagen

Department of Interactive Systems
Centre for Mathematics and Computer Science (CWI)
P.O.Box 4079, 1009 AB Amsterdam, The Netherlands

Abstract: This paper presents the use of ADDL (Artifact and Design Description Language) for design process representation in an intelligent CAD system. ADDL is a special purpose programming language which conceived at the Centre for Mathematics and Computer Science (CWI). It is now being further developed both at the University of Tokyo and at CWI. ADDL is designed for implementing systems which assist a designer in performing a design task. It allows dynamic design object descriptions and a flexible design process representation. In this paper we briefly introduce design object representation in ADDL and elaborate on design process representation. In particular, we pay attention to the multi-world mechanism and the scenario interpreter. In the last section the actual implementation of ADDL on top of Smalltalk-80† is discussed.
Keywords: Knowledge Representation, Object-Oriented Programming, Logic Programming, Intelligent CAD, Multiple Worlds

1. Introduction

Conventional CAD systems, for historical reasons, are centered around a graphical representation of design objects. They are merely drafting systems used in combination with independently operating calculation applications. Designing however, is more than generating drawings of an artifact and testing these through calculation programs. Designing is a stepwise refinement process which takes a rough model of the design object and generates a more detailed model of the design object through a series of design actions [1, 8]. A drawing is more an interpretation of the design object model in a certain context than a description of the design object model. An intelligent CAD system must provide a mechanism to generate several of such interpretations at the same time: a geometric representation, finite element method analyses models, cost analyses models, etc.

†Smalltalk-80 is a registered Trade Mark of Xerox Corp.

An Intelligent CAD system with the above mentioned properties contains a central design object description which we call the *meta-model* [9]. A meta-model is a context free description of a design object. During the course of a design process, several models are derived from this meta-model. These derived models are evaluated in a certain context (a *world* in ADDL) in which case new information is obtained through this evaluation. The meta-model is then extended with the newly derived information, resulting in a more precise design object description. Deriving a model from a meta-model by interpreting it in the context of a world, and updating it with the new information, is called a design step. A design process consists of a number of design steps in a certain order [11].

Another important property of an Intelligent CAD system is the possibility to create multiple views of a design object in parallel. Such a system behaves like several expert systems working on a design object concurrently. We call these expert systems *application modules*. We make a distinction between application modules that work independently from each other and dependent application modules. The former may result in design object descriptions that are different. The latter always produce **one** unique design object description [11].

To implement an Intelligent CAD system which uses a meta-model and a design process model based on stepwise refinement, a special programming language is needed. Such a language allows flexible design object descriptions and means to describe different design steps. It must support a multi-world mechanism to represent dependent and independent design object representations as well. A language fulfilling these requirements is ADDL (Artifact and Design Description Language).[1] ADDL, developed at CWI, is designed for the implementation of the IIICAD (Intelligent Integrated Interactive Computer Aided Design) system [10, 13]. The IIICAD system is a generic system, to be used in arbitrary domains of design. We focus our attention, however, on the fields of architecture and mechanical engineering. In these areas we have the expertise to actually build a prototype system which possesses the functionality described above.

In the rest of this paper we present ADDL, giving a short introduction to its constructs for design object description, and elaborate on its mechanisms for design process representation.[2] In the last section we introduce the ADDL compiler and scenario interpreter. An example of the use of ADDL is given as well.

2. Design Object Representation in ADDL

Constructs for describing a design object in ADDL combine features of object-oriented and logic programming systems. In ADDL we are able to describe both declarative and procedural properties of an artifact. Functions denote procedural aspects while predicates and rules denote the more declarative aspects of a design object. The core of the design object representation in ADDL is a deductive database which we call a *facts-base*. A facts-base involves objects and

---

[1] The language was formerly called IDDL (Integrated Data Description Language). Currently, there are two versions under development, one at the University of Tokyo and one at CWI. To avoid confusion and because the name suits better, we decided to change the language's name to ADDL.

[2] Throughout the paper we will present the fundamental differences between ADDL and IDDL in footnotes like this one.

relationships between objects. In this chapter we show the essential elements of ADDL concerning the facts-base, but avoid details and exceptions.

Before we can describe the facts-base we must introduce the notion of objects in ADDL. All data structures in ADDL, as in other object-oriented languages, are objects. However, we make a distinction between *primitive* objects and *composite* objects. Primitive and composite objects are treated the same in ADDL, except that they are different in their internal structures. Primitive objects are the basic data types of ADDL. They are its building blocks. A primitive object has a type and a value. Examples of primitive object types are: Integer, Real, Character, String, etc. Objects which have more than a type and value own an internal structure; they are called composite objects. The internal structure of a composite object is made of *attributes, functions,* and *constraints.* Composite objects are referred to by their (unique) names. The set of all composite objects is the *objects-base.*

A facts-base is used to describe relationships between objects and consists of *facts.* A fact is a predicate symbol followed by a list of *constants,* separated by commas, e.g. supports(leg, table). A predicate symbol defines a relationship between its constants. A constant is an object's name. Composite objects' names are symbols that start with a lowercase character. The name of a primitive object is its value. For instance, consider the fact numberOfLegs(table, 4). In this example the symbol table is a composite object's name and 4 is a primitive object of type Integer and value 4.

In the current implementation of ADDL predicates can have only constants for their arguments. The next implementation, however, will allow predicates to have functions for their arguments as well. These functions behave like messages, sent to an object. A function call is depicted as follows: an object, for which the function is called, a function name, i.e. a colon followed by a symbol starting with a lowercase character, and an (optional) list of constants separated by commas in square brackets, i.e. anObject :aFunction[aConstant]. An example of a predicate containing a function is: greater(table :numberOfLegs, 4). The object table, in this example, has an attribute named numberOfLegs which can be accessed through a function :numberOfLegs. The entire predicate expresses the fact that there exists an object, table, which has an attribute numberOfLegs, and the value of that attribute should be greater than four, i.e. it can be seen as a constraint. Functions do not only denote attribute values of objects. They may occur as a function declaration in the object's internal structure as well. A function returns a primitive object's value, either an attribute value or a computed value. Hence, there is no difference in behaviour between a function call which returns an attribute value and a call which returns a computed value, when you regard the object's internal structure as a black box.

There are a number of predicate symbols in ADDL which have a special meaning. These built-in predicates describe the structure of the objects-base. First of all, a composite object is instantiated by asserting a one-placed predicate to a facts-base. The predicate symbol denotes the type of an object and its argument its name. An example of an instantiation is the assertion of table(kitchenTable), where table denotes an object's type and kitchenTable an object's name.[3] In ADDL there is a catalogue of *prototype* definitions. When a composite object is instantiated, the prototype definition which corresponds to the object's type is looked up in this catalogue. A

---

[3] In IDDL, the built-in predicate makeObject is used. E.g. makeObject(Table, kitchenTable).

copy of the prototype is made for the object's internal structure. The notion of prototypes in ADDL is similar to that of classes in Smalltalk-80 [4, 5]. However, in ADDL the internal structure of an object may change during its life-time without affecting its prototype definition, i.e. attributes, functions, or constraints may be added or removed to/from an object. This is contrary to Smalltalk-80 where the methods and instance variables belonging to a certain instance cannot be extended without affecting all other instances of the same class.

So far, we discussed built-in predicates for instantiation of individual composite objects. An artifact representation consists of a set of composite objects which together form a detailed description of the design object. In ADDL, the decomposition of a design object into its parts is achieved through a built-in predicate hasPart. It is a two-placed predicate symbol denoting that the second argument is a part of the first, e.g. hasPart(table,leg). The part-of hierarchy of a design object may be several levels deep. If some information is requested from an object, (we shall describe how this is done in the next section), and the object is unable to provide it, then the request is delegated to its parts, as described by the part-of hierarchy, in order to obtain an answer.

In ADDL there is an is-a hierarchy as well. It defines a prototype hierarchy similar to the class hierarchy in Smalltalk-80. It is realised by a two-placed built-in predicate symbol, isA. The isA predicate specifies that all objects which belong to the type of its first argument, are specializations of the type of its second argument. Objects of the first type inherit properties of the second type. An example is: isA(table, pieceOfFurniture); objects of type table inherit properties that are defined for objects of type pieceOfFurniture. A more detailed description of the delegation and inheritance mechanisms can be found in [12]. In the same article a more elaborated description of ADDL concerning the design object representation can be found.

Another built-in predicate is used to structure the facts-base. It behaves like a composite object definition predicate we have introduced above. It defines a special type of composite object, namely a *world*. A world is a facts-base in itself. It contains predicates and it has an objects-base associated with it. A world may contain other world definitions and so on. Worlds partition the facts-base in a straightforward manner. So if we assert world(table) to the facts-base (which itself is a world), we create a new facts-base with the name table to which we can assert new facts. The use of worlds will be discussed further in the next section.

3. Scenarios

In the previous section we discussed the constructs in ADDL which concern the representation of a design object, the so called meta-model. In order to build a meta-model, ADDL provides mechanisms to reason about the current state of a design object and to generate new information from it. The construct to achieve this is called a *scenario*. A scenario contains both procedural and declarative knowledge in order to evaluate the design object model in a certain context and to derive new information. A scenario has a world (facts-base) associated with it, a set of functions (optional) and a number of IF-THEN rules [2]. A sequence of scenario calls corresponds to a series of design steps. A scenario is invoked with a world associated with it. Execution of a scenario involves firing of rules and calling of functions. A scenario remains active until a certain stop condition has been reached. Finally, control is given back to the caller of the scenario. The results of the call are registered in the world associated with the called scenario and these are merged into the world of the caller, if and only if the two worlds are consistent with each other. Let us look now briefly at the syntax of scenarios and then concentrate on how a scenario interacts with its world when it is executed.

3.1. Syntax of scenarios

A scenario has a name, a rule selection mechanism in parentheses (optional), a set of function declarations (optional) starting with the keyword FUNCTIONS, and a number of IF-THEN rules starting with the keyword RULES. Thus, a scenario may look like:

```
scenarioName ( ruleSelectionMethod )
FUNCTIONS
   :f = { functionBody }
   :h[ argumentList ] = { functionBody }
RULES
   IF condition THEN action
```

A scenario is just another composite object and thus has a unique name. A rule selection method determines in which order the rules are fired. We shall come to this later when we deal with the execution of scenarios. A function has a name, i.e. a colon followed by a symbol starting with a lowercase character, and a number (may be zero) of arguments. An argument list is a sequence of variables (symbols starting with a capital) separated by commas. A function body consists of a number (may be zero) of temporary assignments separated by semicolons, followed by a return expression. All variables that occur in a function body are automatically defined as local variables within that function. An example of a function declaration is:

```
:area[ R ] = RR := R * R ; 3.14 * RR .
```

Function calls may occur within a function body as well, however, we do not allow recursion. Note that such a function call already appeared in the previous example. The function '*' is called upon R with R as its argument. Strictly speaking the expression should have read something like: R :multiplyWith[R], but for convenience we adopted a more readable notation.[4]

An IF-THEN rule describes a design action. The condition part is an evaluation of a certain expression in which some variables may be bound from, for instance, a facts-base (an alternative way to bind a variable is via a user-interface call). A condition, however, never changes the state of a facts-base, it has a so called read-only permission. It is justified to compare a condition with a query to a database, it examines the current state of the design object description. The action part adds new information to a design object description either by asserting new facts or by assigning values to attributes.

Both conditions and actions are called *formulae*. In order to define what a formula is we must define *terms* and *predicates* first. A term is defined as follows:

---

[4] A similar strategy is used in Smalltalk-80 for the arithmetic and boolean operators.

1. A variable is a term, e.g. X, Y, ATable.
2. A constant is a term, e.g. a, b, aTable, 4, 3.14, 'hello world', etc.
3. A function call is a term, e.g. aTable :area[X], X :f.[5]
4. If t is a term, then (t) is a term.

A *ground term* is a term not containing variables. If p is an n-placed predicate symbol (a symbol starting with a lowercase character) and $t_1, \ldots, t_n$ are terms, then $p(t_1, \ldots, t_n)$ is a predicate. A *predicate constant* is a zero-placed predicate. In ADDL the only predicate constants are, T (true), F (false), succeed, and fail. A formula is defined as follows:

1. A predicate is a formula.
2. If p is a predicate, then %p is a formula.
3. If p and q are formulae and ⊗ is a binary connective, then p ⊗ q is a formula.
4. If p is a formula, then ~p is a formula.
5. If p is a formula, then (p) is a formula.

So far, we defined the unary operators % (the *unknown* operator) and the unary connective ~ (negation). The binary connectives are & (logical and) and | (logical or). Hence, a valid formula is for instance: p(a) & ~ (%q(a) | r(b)).

Basically, ADDL is equipped with two-valued logic. Therefore, if a fact is queried from a facts-base, the result is either 'true' or 'false'. A fact may appear either positively or negatively, e.g. p or ~p. Similarly, we can pose a query as p or ~p which evaluate to 'true' if respectively p or ~p occurs in the facts-base. Furthermore, a facts-base in ADDL behaves according to the ''closed world assumption'' [7]. Thus, the query of a fact fails either because the negation of that fact appears in a facts-base, or it was not present at all (unknown). In some situations, however, we want to know explicitly whether a fact is unknown or not. For that case, ADDL provides the unknown operator, %. Its truth table is built up as follows; let p be a predicate, then evaluate

%p to { true — if neither p nor ~p can be found in the facts-base.
false — if either p or ~p appear in the facts-base. }

With the use of % we have implicitly introduced a third truth value, the unknown. The % operator can be used only in condition formulae. It may be used with both constants and variables. The former checks whether a certain predicate is absent in a facts-base. The latter checks whether a set of predicates is absent. For instance, %p(a) succeeds if neither p(a) nor ~p(a) appear in the facts-base. The query %p(X) checks whether the one-placed predicate p or its negation appears in the facts-base with any argument. Note that a binding for X cannot possibly occur, since X will only be bound when there exists a unary predicate p. However, in that case the query fails and it, therefore, produces no results.

---

[5] We do not allow a nesting of function calls, thus only constants and variables may appear in a function call.

3.2. Evaluation of a condition formula

As mentioned before, the predicates in the condition formulae evaluate the current state of a design object description. For condition formulae we distinguish the predicate constants, T and F, and three kinds of predicates: (1) queries to a facts-base, (2) built-in predicates, and (3) user interface queries. The predicate constants obviously evaluate to the values they represent. When a predicate from the first category occurs in a condition formula, the facts-base is searched for a predicate with the same predicate symbol and the same arity. If such a predicate is found, its arguments are matched against those of the predicate appearing in the condition predicate (the actual unification mechanism is explained below). Such a query evaluates to true if the match succeeds. Predicates that do not belong to the second and third category automatically belong to the first.

Built-in predicates perform a test on their arguments. Examples of these are: equals, greater, smaller, known, etc. The terms of such a predicate are evaluated and compared with each other according to the type of predicate symbol. For instance, the built-in predicate equals evaluates to true in case of equals(a, a), and it evaluates to false in case of equals(a, b). These are of course trivial examples. The evaluation is less straightforward, if the terms are variables or function calls, e.g. equals(table :numberOfLegs, 4). The predicate is evaluated with the return value of the invoked function. However, when a function returns 'nil', for instance when an attribute value is not (yet) defined, the entire predicate evaluates to 'unknown', and hence the predicate fails. Furthermore, if we pose the query %equals(table :numberOfLegs, 4), the predicate will return true if the function returns nil, otherwise it will fail. Moreover, the query known(table :numberOfLegs) succeeds if the attribute numberOfLegs is already determined.

Predicates from the third category behave similar to those of the first category, but with the difference that the user is queried instead of a facts-base.

A formula is evaluated from left to right. The operator % has the highest precedence, followed by ~, while & and | have the lowest precedence. Variables that occur in a rule are local to that rule. They are bound during the evaluation of the condition. If there are still unbound variables after an evaluation, the rule fails.

Presently, ADDL uses a very simple unification algorithm [6] to bind variables. The predicates occurring in a formula are processed one by one, gradually reducing the number of possible bindings. In short the algorithm behaves as follows. The first predicate is unified with the facts-base that is associated with the scenario's world. The unification results in an *instantiation pair list* containing all possible bindings. The same happens with the second predicate resulting in another instantiation pair list. The two lists are compared, conflicting bindings are removed and the two lists are merged into one. Then the next predicate is evaluated, its instantiation pair list is merged with the result list, and so on. The algorithm is best described with an example. Suppose a facts-base contains the following facts:

```
p(a).       p(b).       p(c).
q(a, b).    q(a, c).    q(b, c).
r(b, c, b).
```

Now, the formula,

p(X) & q(X, Y) & r(X, Y, Z)

is posed to this facts-base. For the first predicate an instantiation pair list, {X → a;b;c} is found. Unification of the second predicate generates {X,Y → a,b;a,c;b,c}. A merge of the two lists results in: {X → a;b. X,Y → a,b;a,c;b,c}. Note that c is removed from the first list. The third predicate produces {X,Y,Z → b,c,b}. The resulting instantiation pair list is then {X → b. X,Y → b,c. X,Y,Z → b,c,b}. The final result of the unification is that X is bound to b, Y to c, and Z to b. Furthermore, the condition succeeds.

3.3. Evaluation of an action formula

An action formula adds new information to a design object description. All variables that occur in an action formula must be bound at the time the action is evaluated. If there still exists an unbound variable, the rule will produce a run-time error. Similar to condition formulae, there are several categories of predicates available for an action formula. We distinguish four kinds of predicates: (1) assertions to a facts-base, (2) assignments to attributes, (3) scenario calls, and (4) control predicates.

A predicate symbol which is not an assignment, a scenario call, or a control predicate is automatically regarded as an assertion to the scenario's world. The terms of a predicate are evaluated before assertion, i.e. the functions are evaluated prior to the assertion. So, the assertion of greaterThan(table :numberOfLegs, 4) will result in a predicate with two constant terms.

Assignments in ADDL are basically allocations of attribute values. An assignment is made with the built-in predicate gets. It is a two-placed predicate, which assigns the value of the second argument to the first. The first term is a function expression, i.e. a reference to an object's attribute. The second term produces the value to be assigned, e.g. gets(table :numberOfLegs, 4). Only uninstantiated attributes may be assigned. Therefore, an assignment fails if an attribute is already determined. Besides, an assignment predicate fails when its first term is not a function expression, or when the function is not a reference to an object's attribute.

ADDL provides a mechanism which allows the user to invoke a scenario from a rule, and hence, to produce several levels of scenarios. We call a scenario from which a scenario is invoked, the *parent* scenario. A built-in predicate, use, is employed in ADDL to call a scenario.[6] Its first argument is a scenario name. Further arguments of the predicate are references to worlds. These worlds are associated with the scenario during its activity. A scenario remains active until a certain stop condition is encountered (represented by control predicates, which we will discuss below). An example of a scenario call is: use(designTable, tableWorld). When there are no world references in a scenario call, the invoked scenario will use the worlds of the parent scenario. The results of a scenario are checked for consistency with the worlds of the parent scenario after successful termination. These results, i.e. assignments and assertions, are consistent if the assertions do not violate constraints in the parent scenario's worlds and if the negations of the assertions are absent in the parent scenario's worlds.

---

[6] In IDDL, the built-in predicate select is used.

The following control predicates can be found in action formulae: succeed, fail, and directive. The first is a zero-placed predicate which causes a scenario to succeed immediately, and returns control to the parent scenario. The second, also zero-placed, behaves the opposite way. It causes a scenario to fail with loss of all results. The third one-placed built-in predicate is used to change the current rule selection mechanism dynamically. For instance, when directive(recency) is encountered, from now on the rule selection method recency will be applied by the scenario interpreter until another directive is encountered.

3.4. Multiple worlds

The multi-world mechanism in ADDL allows for invoking multiple scenarios concurrently. This implies that there are more than one scenario active at the same time. Moreover, it means that multiple worlds are accessed concurrently. Two additional binary connectives are employed to invoke the multi-world mechanism. These are: && (parallel and) and || (parallel or). The former is used to create dependent worlds. This dependency implies that the scenarios' worlds have to be consistent with each other, at the time all dependent scenarios have terminated. The following action formula is an example of the invocation of two dependent scenarios,

use(scenario1, world1) && use(scenario2, world2).

Dependent worlds in ADDL allow a user to generate different interpretations of a meta-model concurrently.

The || connective is used to create independent worlds. Independent worlds, in contrast with dependent worlds, do not necessarily have to be consistent with each other. Inconsistent independent worlds result in distinct design object descriptions. Independent worlds provide a user with a mechanism to model distinct design solutions. An example of the invocation of independent worlds is:

use(scenario1, world1) || use(scenario2, world1).

3.5. The modal operator

Designers are used to reason about uncertain facts. ADDL is therefore equipped with a modal operator P. The operator P is a unary operator. Syntactically it is used similar to the operator %. However, P appears only in action formulae. The assertion of a predicate which is preceded by P results in an *assumed* fact [3]. Normal assertions to a facts-base are permanent. It may be convenient, however, to assert a fact which is not absolutely certain, as an assumed fact. Such a fact may in a later stage of the design process be withdrawn, or become certain. Such an assertion is, for instance, P shape(table, round).

A fact which is derived from an assumed fact is an assumed fact as well. A fact is derived from all facts that appeared in the condition of the IF-THEN which asserted it. A derived fact has a list of dependencies associated with it. This list contains the assumed facts from which it is derived (only one level deep). At the time an assumed fact becomes certain, it is removed from the dependency lists in which it occurs. Facts, that depend only on this fact, become certain as well. An assumed fact cannot cause an inconsistency. If it does, it and all facts depending on it, are removed from the facts-base.

4. The Scenario Interpreter

In this section we illustrate the execution of ADDL scenarios with an example of the design of a part of a table. The encoded scenarios are shown in Fig. 1. to Fig. 3.

When the scenario in Fig. 1 is executed, the first rule is selected according to the specified rule selection method. In this case it is the default rule selection method, i.e. straightforward top-down selection, and rule #1 will therefore be selected. Variables are matched against the world associated with scenario tableArea. This world is passed with the scenario call that caused the activation of the scenario. Another possibility is that the world is empty, because tableArea is the top-level scenario. In case of rule #1, if there does not exist a predicate table in the world, the query %table(X) will evaluate to true and the predicate fail will cause the execution of the scenario to terminate unsuccessfully.

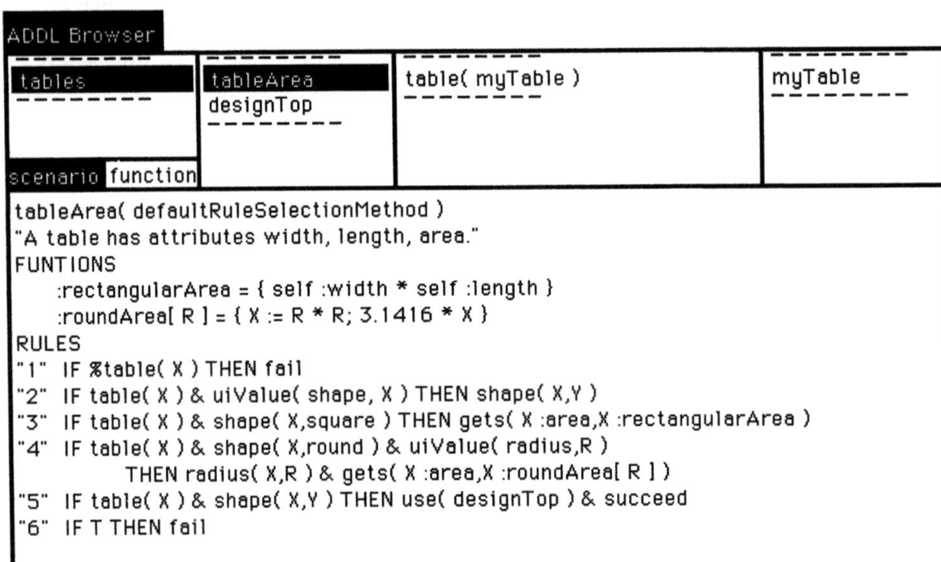

Fig. 1. Scenario tableArea before execution

The existence of any one-placed predicate table will cause rule #1 to fail and the next rule to be selected. For this scenario to be of any use, there must exist a table-predicate, for instance table(myTable). If the presence of more than a single table is allowed, the conditions must be extended with extra identification queries. For the sake of simplicity, let us presume that myTable is the only table present in the world.

The condition of rule #2 consists of a conjunction formed by the conditional and. The first part of the condition will evaluate to true with X bounded to myTable. The second part is a user-interface request, i.e. the user is asked to enter the shape of the table. If the user has entered a non-empty string, say round, this query will succeed with Y bounded to round. As a result the predicate shape(myTable, round) will be asserted to a temporary facts-base. Rule #3 and #4 compute the table area, depending on the table shape. In the case of a round table, rule #3 will not match but rule #4 will. Being a round table the user is asked for the radius, say fifty, the predicate radius(myTable, 50) is asserted and the result of the function call myTable :roundArea[50] is assigned to the attribute area of myTable.

Rule #5 will match, resulting in a call to the scenario designTop. If designTop terminates successfully, its results will be added to the facts-base of tableArea and the predicate succeed will cause the execution of tableArea to terminate successfully. Likewise, the results of tableArea will be added to the facts-base of its parent scenario. Rule #6 will only be fired if rule #5 did not succeed, i.e. it did not match because one of the queries failed or the scenario call failed. Rule #6 will always succeed because T evaluates to true and fail will cause the execution of tableArea to terminate unsuccessfully.

Because no worlds are explicitly mentioned in the scenario-call in rule #5 of scenario tableArea, the world of scenario designTop (Fig. 2.) is identical to the world of its parent scenario. Therefore, its first rule will match, the predicates top(roundTop) and hasPart(myTable, roundTop) will be asserted and control will be returned to tableArea. As a result of the assertion of top(roundTop), the object roundTop will be added to the objects-base because top is a prototype. The results of designTop will be added to the facts-base of tableArea.

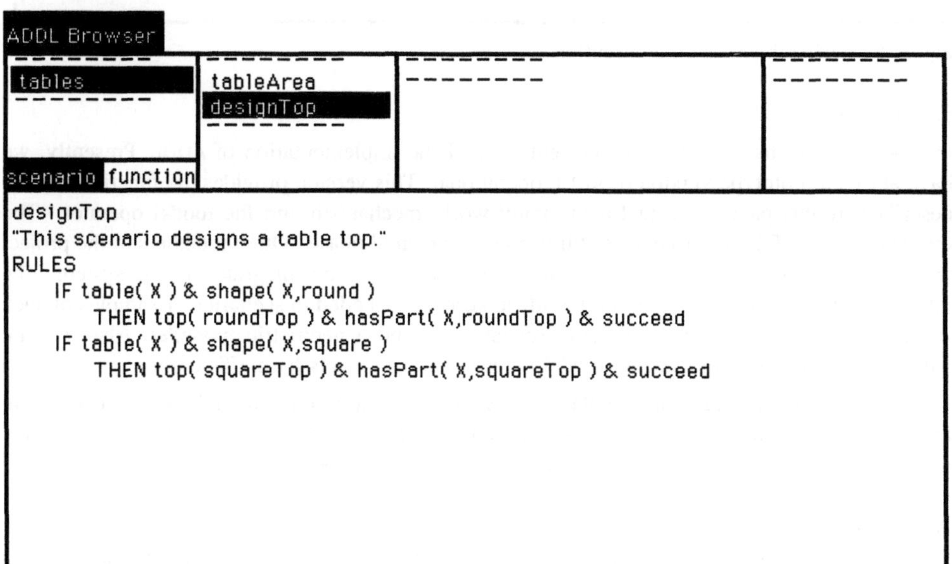

Fig. 2. Scenario designTop

In Fig. 3. the results of the execution of tableArea are visible. The facts-base contains five facts and the objects-base contains two objects.

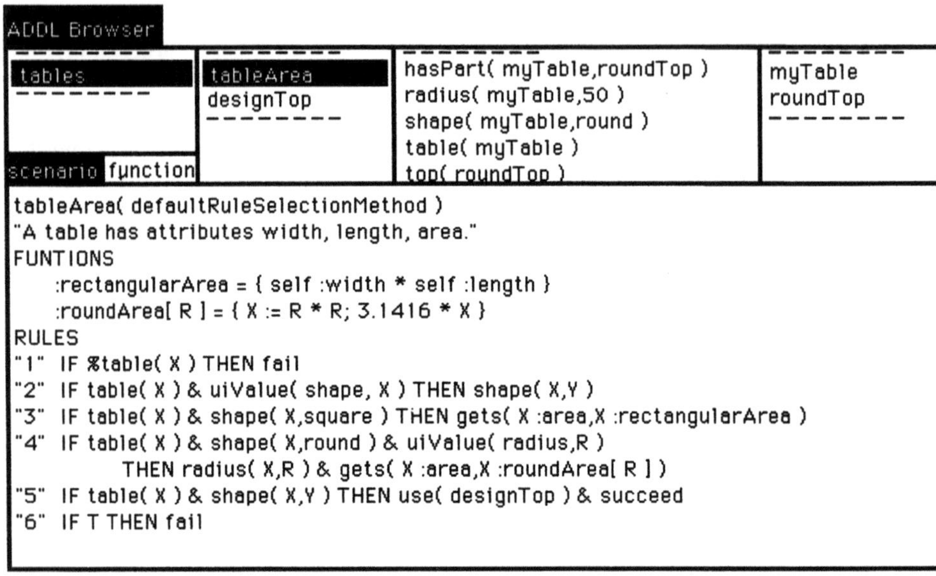

Fig. 3. Scenario tableArea after execution

5. Conclusions

In this paper, we have showed the current state of the implementation of ADDL. Presently, we have at CWI a prototype version of ADDL operational. This version provides the functionality, as described by this paper, except for the multi-world mechanism and the modal operator. The implementation of these concepts is still in an experimental phase. Other members of the project team are working on application modules which can easily be integrated in the system. The interface between these modules and the ADDL kernel is encoded in scenarios. Examples of such application modules are an intelligent user-interface and a geometric modeller. Also, we are working on an underlying database model to represent the objects-base efficiently.

The next phase of the implementation of ADDL will incorporate the multi-world mechanism and the modal operators into the prototype system. Furthermore, an extensive example in the domain of architectural design (floor planning of a house) will be encoded in ADDL.

Acknowledgements

The authors wish to thank Tetsuo Tomiyama, Varol Akman, Farhad Arbab, Jan Rogier and Sabine Eisses for the many fruitful discussions, which have contributed a lot to this paper. Furthermore, we would like to thank Monique Megens whose pioneering work on the implementation of ADDL is invaluable.

References

1. V. Akman, P.J.W. ten Hagen, J. Rogier, and P.J. Veerkamp, "Knowledge Engineering in Design," *Knowledge-Based Systems*, vol. 1, no. 2, pp. 67-77, 1988.
2. R. Davis and J. King, "An Overview of Production Systems," in *Machine Intelligence*, ed. E.W. Elcock and Donald Michie, pp. 300-332, Ellis Horwood Ltd., Chichester, 1977.
3. J. De Kleer, "An Assumption Based TMS," *Artificial Intelligence*, vol. 28, pp. 127-162, 1986.
4. A. Goldberg and D. Robson, *Smalltalk-80: The Language and its Implementation,* Addison-Wesley, Reading, 1983.
5. A. Goldberg, *Smalltalk-80: The Interactive Programming Environment,* Addison-Wesley, Reading, 1984.
6. J.W. Lloyd, *Foundations of Logic Programming,* Springer-Verlag, Berlin, 1987.
7. R. Reiter, "Towards a Logical Reconstruction of Relational Database Theory," in *On Conceptual Modelling*, ed. M.L. Brodie, J. Mylopoulos, and J.W. Schmidt, pp. 191-233, Springer-Verlag, New York, 1984.
8. P.J.W. ten Hagen, J. Rogier, and P.J. Veerkamp, "An Environment for Knowledge Representation for Design," in *Proceedings of Civil Engineering Expert Systems*, Madrid, 1989.
9. T. Tomiyama and P.J.W. ten Hagen, "Representing Knowledge in Two Distinct Descriptions: Extensional vs. Intensional," Report CS-R8728, Centre for Mathematics and Computer Science, Amsterdam, 1987.
10. T. Tomiyama and P.J.W. ten Hagen, "Organization of Design Knowledge in an Intelligent CAD Environment," in *Expert Systems in Computer Aided Design*, ed. J. Gero, pp. 119-147, North-Holland, Amsterdam, 1987.
11. P.J. Veerkamp, "Multiple Worlds in an Intelligent CAD System," in *Intelligent CAD, I*, ed. H. Yoshikawa and D. Gossard, pp. 77-88, North-Holland, Amsterdam, 1989.
12. P.J. Veerkamp, V. Akman, P. Bernus, and P.J.W. ten Hagen, "IDDL: A Language for Intelligent Interactive Integrated CAD Systems," in *Intelligent CAD Systems II – Implementational Issues*, ed. V. Akman, P.J.W ten Hagen and P.J. Veerkamp, pp. 58-74, Springer-Verlag, Berlin, 1989.
13. Bart Veth, "An Integrated Data Description Language for Coding Design Knowledge," in *Intelligent CAD Systems I – Theoretical and Methodological Aspects*, ed. P.J.W ten Hagen and T. Tomiyama, pp. 295-313, Springer-Verlag, Berlin, 1987.

11

Discussion Session: Languages

Farhad Arbab

1. Papers

The second day of the workshop started with a session on Languages for Intelligent CAD systems. Three papers were presented in this session, one on ADDL, one on IDDL and one on GRAFLOG. The first paper was presented by Paul Veerkamp, and emphasized the representation of design processes in ADDL. The use of formal logic in ADDL and the concept of scenarios seemed to draw attention.

In the discussion of this paper, Shenton asked about the advantages of ADDL over Prolog. Veerkamp explained that the object-orientedness of ADDL supports class hierarchies and message passing. On the logical side, the concept of unknown and the modal operators make the ADDL paradigm different than that of Prolog. Aart Bijl asked if this meant that ADDL uses three-valued logic. Veerkamp replied no. He explained that ADDL does not use the closed world assumption. As a result, it is possible to derive the conclusion that the truth value of some formula is not known in some world. Bijl asked for more elaboration on how this 'not-known' is used. Because of time limitations, this question was postponed to be picked up during the general discussion session.

MacCallum raised the issue of functions between two objects, i.e., which object is the function linked to. For example, he asked, how is the relation between two legs of a table represented? Veerkamp stated that such a function will not be associated with either leg of the table. Rather, it will be linked to the 'assembly object' that contains the two legs. This is simply stated as a fact in the facts base.

El Dahshan asked what, exactly, a scenario is in ADDL. Veerkamp stated that it is a set of transactions. El Dahshan raised the issue of contradictory facts and how they are dealt with in ADDL. This topic was also left for the general discussion.

In the second paper presentation, Luis Pineda described a representation scheme based on formal logic for capturing CAD knowledge. He demonstrated the use of this scheme in dealing with the semantic contents of the dialogue between a system and a designer. He described GRAFLOG, a program implemented in Prolog, as a system that uses this scheme, and briefly explained its syntax and semantics. His examples in the presentation emphasized acquisition and use of higher-level semantics of graphical constructs to support a more intelligent interaction with a designer. The representation scheme, however, is not meant to support a "user interface" exclusively. It is proposed as a formalism for capturing CAD knowledge, in general.

In the question period following this presentation, the user interface issue was raised and Pineda was asked about the interpretation of a user's actions, e.g., pointing to lines and other graphic entities. For simple cases, it is possible to come up with good heuristic rules, e.g., the common gravity of pointer rule. But, what more can be done? He stated that correct understanding of users' intentions through interpretation of their actions is a general problem in all systems. He sees no additional complications resulting from the use of his formalism to represent the acquired knowledge.

Daan Otten asked if his system knew what, e.g., 'parallel' really meant, and if such knowledge was actually expressed declaratively. Pineda stated that his system indeed knows what 'parallel' is. He added that in this case, it is not stated declaratively, instead, it is a derived concept. Joubert asked if GRAFLOG was meant as a stand-alone system or a front-end to some other system. Pineda responded that it is now a front end.

MacCallum pointed out that the difference between the graphical model and its geometry is not very clear in this paper. Every action seems to be stated in terms of the underlying geometry, and everything seems to be evaluated sequentially. Pineda stated that a different paper by him gives a more detailed explanation of the system and he could not elaborate on this issue further, because of time constraints. Shenton asked if he had considered solids, and Pineda replied no. The issue of graphical interaction was raised again and Pineda was asked about dealing with overlapping lines on the screen. This topic was also left for the general discussion.

In the third paper presentation, Tomiyama described their experience with development and use of IDDL at the University of Tokyo, and discussed examples of design knowledge representation in this language. The IDDL work at the University of Tokyo is now a parallel effort with the ongoing project at CWI. It was clear that the two implementations are different, and it seemed there are even some conceptual differences between the two projects.

Following the presentation, Barthès remarked that he felt there was a confusion about the notions of classes, prototypes, inheritance, and delegation. He added that in his opinion IDDL contains unnecessary things, and much is missing in its present description as a language. He then asked if the IDDL representation of a design process is dynamic. For instance, is it possible to add a new class during a design session. Tomiyama stated that it is possible to add new concepts in scenarios.

Treur raised the issue of accessibility and the modal operators. Tomiyama stated that the use of modal logic operators in IDDL must be elaborated more clearly. Treur then commented that there are three types of knowledge used in design: design requirements, object representation, and strategies. He asked where in IDDL each is represented. This issue was left for the general discussion session. Bernus asked where modal operators could be used in IDDL. Tomiyama replied that they can be used in both queries and assertions.

2. General Discussion

The apparent difference between the two versions of IDDL, viz. ADDL and IDDL was raised and discussed. Mac an Airchnnigh suggested that different names be given to the two languages developed in parallel at CWI and at the University of Tokyo, to avoid confusion.[1]

Paul Veerkamp elaborated on the use of the unknown operator in ADDL. In addition to positive facts, negative facts can be asserted in the fact base of ADDL. The underlying model used here *is* based on the closed world assumption. This means that the truth value of a query is true only if it is found to be true in the facts base, and it is false, otherwise. There are cases, however, where it is required to know if a query is false because its negation exists in a facts base, or simply because we fail to prove it. A special operator, %, is made available in ADDL, called the unknown operator, for this purpose. Addition of this unknown operator effectively lifts the closed world assumption and implicitly introduces a third truth value. He also discussed the multiple worlds feature of ADDL.

Tomiyama elaborated on the scenarios in IDDL and how they incorporate different types of design knowledge. Pineda discussed how ambiguity resolution rules are used in GRAFLOG.

---

[1] Editor's comment: The original paper of Veerkamp et al. dealt with IDDL. In the revised version the name was altered to ADDL.

Geometric Reasoning

12

GARE: Geometric Analysis and Reasoning Engine

Rajiv S. Desai, Rajkumar S. Doshi, and Raymond K. Lam

Artificial Intelligence Group
Jet Propulsion Laboratory
California Institute of Technology
Pasadena, CA 91109, USA

Abstract: A geometric analysis and reasoning engine (GARE) has been developed for geometric modeling and geometric problem solving for robot task planning application. Geometric modeling and geometric reasoning are crucial components of the design and operation phases of the modern automated systems. Intelligent Computer Aided Design (ICAD) systems should not only allow for geometric modeling but also provide capabilities for reasoning with the geometry. Many interesting physical analysis problems that arise in various applications such as manufacturing and robotics can be transformed into equivalent geometric problems. The basic concept underlying the design of GARE is that many complex geometric reasoning problems can be solved by combining simple geometric algorithms. GARE consists of a geometric modeling system, a geometric task description system, and a geometric reasoning system. This paper presents an overview of various components of GARE and their integration.
Keywords: Geometric Modeling, Geometric Reasoning, Motion Planning, Task Description Language, Intelligent CAD.

1. Introduction

Computer modeling is a crucial and integral component of the modern automated systems, such as robot systems. A model based automated system must represent the system, environment of the system and, the complex interactions among the components of the system and the environment. The models must reflect the behaviour of the system and that of the environment. In general modeling techniques dependent on the characteristics of the system and the function of that system. Several commercial CAD modeling systems allow sophisticated modeling features. Detailed surveys of geometric modeling systems can be found in [1, 14].

While several very sophisticated geometric modeling systems have been developed there are very few that allow for modeling and analysis of the physical processes. Usually modeling of physical processes is done separate from the geometric modeling. Typical analysis systems include dynamic simulation, finite element analysis, geometric simulation of manufacturing processes [4, 19]. An intelligent computer aided design system (ICAD) should not only allow for geometric modeling but also offer the interfacing systems with the functionalities to reason about the geometry.

The Geometric Analysis and Reasoning Engine (GARE) was motivated by the need for a system for geometric reasoning during development of robot task planning [6] and knowledge based diagnostic systems [9] for NASA's telerobot project. Typical robot tasks involve reasoning about collision free path, finding stable poses of objects, determining assembly sequences etc. An automated robot system may consists of robots, parts, sensors, computer vision systems etc. A typical task may require assembling a set of parts in some prespecified configuration. This can be posed as a geometric planning problem, where the world can be represented by a geometric model, then the problem is equivalent to moving the parts from one geometric configuration to another in that world. However this may require reasoning about several geometric problems, e.g., generation of assembly sequences, planning collision free paths for the robots, determining grasp positions etc. In general such a system must be able to reason about the complex geometric interactions among it's different components, such as, interference among objects, stable positions of objects, kinematic constraints of the system etc. Geometric Analysis and Reasoning Engine (GARE) is a general purpose tool for reasoning about various complex geometric interactions.

2. Geometric Analysis and Reasoning Engine (GARE): An Overview

GARE consists of four major subsystems, namely, geometric modeling system, geometric reasoning system, task description language interpretation module (TDLIM) and GARE/user interface as shown in Fig. 1.

The geometric modeling system is used to create the the model of the environment. Geometric reasoning systems (GRS) is a collection of geometric algorithms, geometric and motion analysis routines, and controls routines. A user or an interfacing system can make a geometric reasoning task specification using the task description language (TDL). TDLIM interprets the TDL commands and passes the commands to appropriate subsystem. Each of these systems are described later in more detail.

The GARE system is based on the idea that many complex geometric problems may be solved by combining solution methods for simpler geometric problems. Consider the task of determining a position for an object such that object is statically stable. The stable poses of an object can be found be found by computing the convex hull of an object and determining the stable faces on the convex hull. This geometric problem can be solved by appropriately combining property computation routines (convex hull, center of mass etc.) and performing a static stability analysis for each pose. Now suppose the problem was to determine stable positions that are graspable by a given robot. This problem can be solved by combining the previous computations with selector function that reduces a set of positions to a set of positions that are graspable. Now, the problem could be further complicated by requiring that the positions should be stable, graspable and reachable by the robot. GARE would apply another

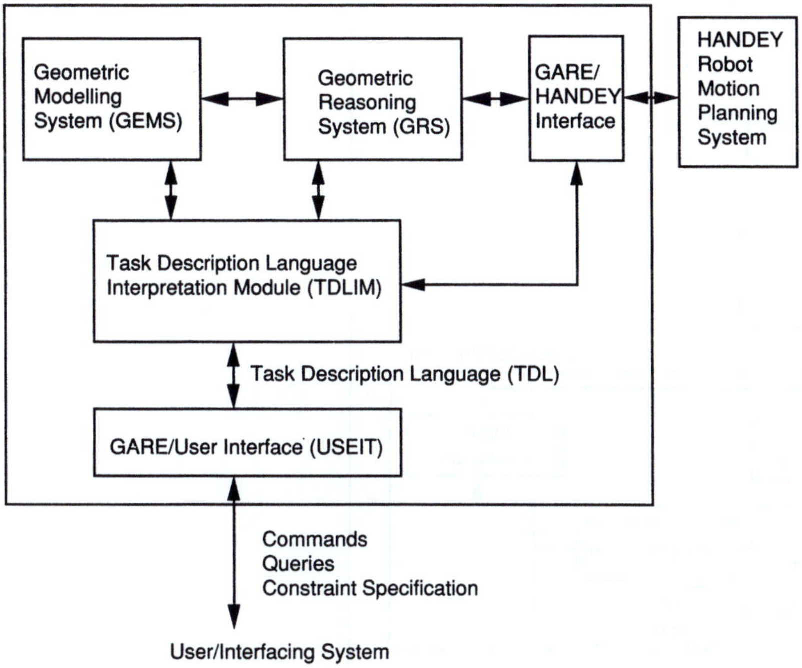

Fig. 1. GARE: Geometric Analysis and Reasoning Engine

selector function to the output of the previous problem that reduces a set of positions to set of positions reachable by a robot.

When a task specification is presented to GARE, the TDLIM parses the input and then performs task translation as shown in Fig. 2. Given a geometric problem like the one described above, GARE attempts to construct a solution plan, which is an ordered list of function calls to various geometric computation modules in the geometric reasoning system, and determines the inputs at each stage in the plan. Solution plans are generated using a rule base. If the inputs are not available, appropriate input constructions functions are invoked.

Input construction is done by determining the availability of the data at correct level of representation (logical, topological or metric). If the data is not available at the correct level of abstraction, data transformation routines are invoked. Sometimes there are some constraints that may need to be applied before inputs are collected. For example, given all the stable positions of an object only the positions that are reachable by a specified robot may be of interest as shown in Fig. 3(a). GARE determines the appropriate selector functions using the selector rule base and selector transition graph (shown in Fig. 3(b)).

Once all the inputs required for each of the function in a solution plan have been constructed, the functions are applied in the order dictated by the plan. If the output of any function call does not conform to the input of the next function call, appropriate data transformation functions are inserted. The rules for ordering the functions and determining

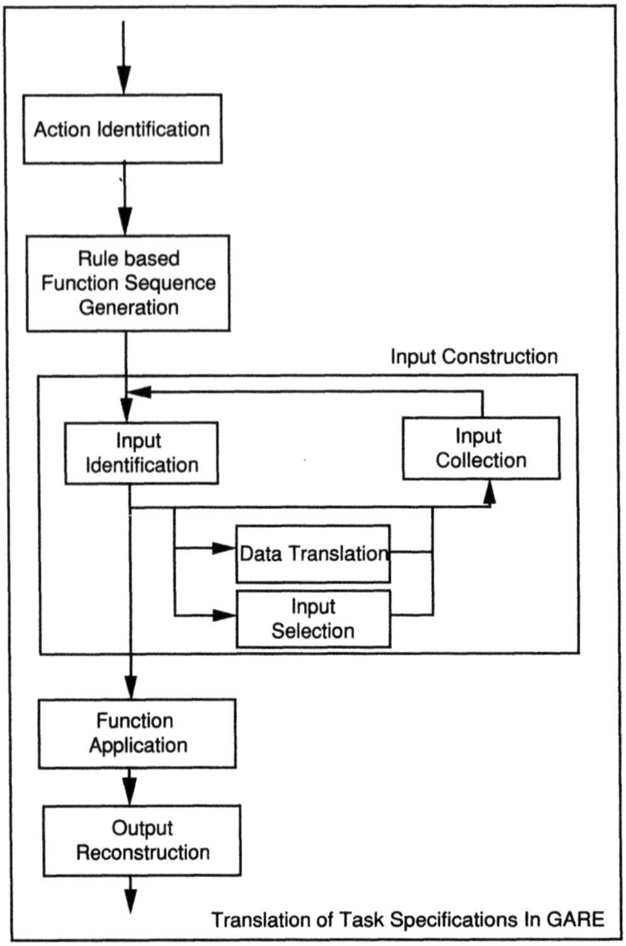

Fig. 2. Translation of task specifications in GARE

appropriate data transformation functions and selector functions are predetermined by a human expert.

Consider the example of a command to find space for an object such that it is stable, graspable and reachable by the robot. Fig. 4 depicts the task translation. The command is parsed by the TDLIM and action is identified to be 'find space' and the constraints are also translated to appropriate level of abstraction. The GARE rule base associates the 'find space' action with the function find--space(face--list1,face--list2) function in GRS. Next GARE checks if any constraints were specified. For each constraint (i.e., stable, graspable and reachable) a function sequence or selector function is determined by constraint analysis in GRS. Now the functions are ordered according to a solution plan and input construction is performed. Finally,

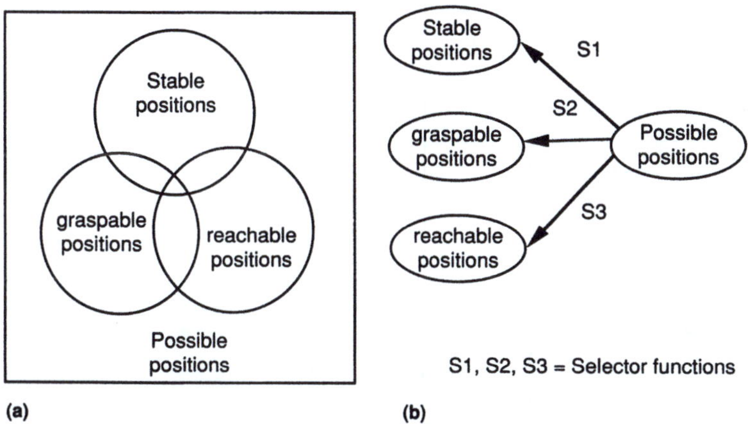

Fig. 3. Selector functions: (a) partitions of positions by constraints (b) selector transition graph

each function is invoked preceded and proceeded by corresponding input construction and output reconstruction routines respectively.

An important component of a modeling system is it's user interface. The interface besides being easy to use must also be able to prompt the user to reasonable and correct inputs. The system must be able to detect incorrect inputs and suggest alternatives. GARE's user interface (USEIT) consist of a graphical input system and a command line interface. A user can not only specify the basic modeling commands but also complex assembly and kinematic relations among objects via this interface. A special feature of the interface is that the input to the modeling system may be provided any of the several levels of abstractions. For example, position of an object may be specified by specifying exact coordinates or it's relative contact conditions with respect to other objects in the environment. GARE internally reduces all specifications to appropriate canonical forms. When specifying some inputs, GARE checks for completeness, correctness, and consistency of the input. For example, if a position and orientation of a block is to be specified and if only a vertex position is specified the system would prompt for more input. On the other hand if redundant or inconsistent inputs are provided, the system would indicate that as well. All inputs specifying the object configurations directly or indirectly are translated into canonical forms consisting of algebraic relations and constraints. For instance, input that a vertex should lie on a face is translated to set of linear algebraic constraints. The consistency of the input is then tested by verifying the consistency of these constraints by simultaneously solving the algebraic inequalities.

Next, we describe the major subsystems of the geometric analysis and reasoning engine.

3. GEMS: Geometric Modeling System

The Geometric Modeling System (GEMS) of GARE is used to create and manipulate the geometric models of objects. GEMS is a simple boundary representation based general purpose 3D object modeling systems. Currently, the modeling is limited to polyhedral objects. GEMS also

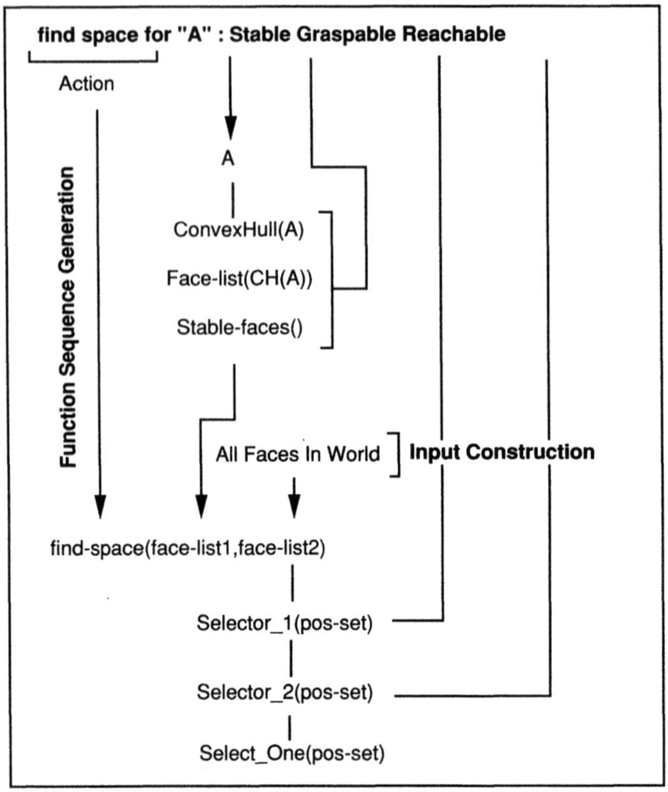

Fig. 4. **TDL task translation for a find space with constraints command**

allows for modeling assemblies and articulate objects (e.g., robot, linkages etc.). Besides modeling mechanical objects such as assemblies and articulate objects, GEMS also allows for modelling two basic types of physical constraints for object models, namely, contact constraints, which are modeled using the concept of contact formations [5], and non-contact constraints such as motion ranges, allowable motion types etc.

3.1. Knowledge representation in GEMS

Various representations schemes are used in solid modeling [13, 15, 16, 22]. GEMS uses a hierarchical representation of objects and constraints. There are three types of representations in GEMS; logical, topological and metric. Consider the assembly depicted in Fig. 5; as shown in Fig. 6, at logical level the assembly is represented as a collection of parts 1, 2 and 3. At the topological level the assembly's structure is represented by an assembly tree, which reflects the order of assembly. Finally, at the metric level the relative transformations among the objects are stored. Object topology in GEMS is represented using the wind-edge data structures [2, 3, 21].

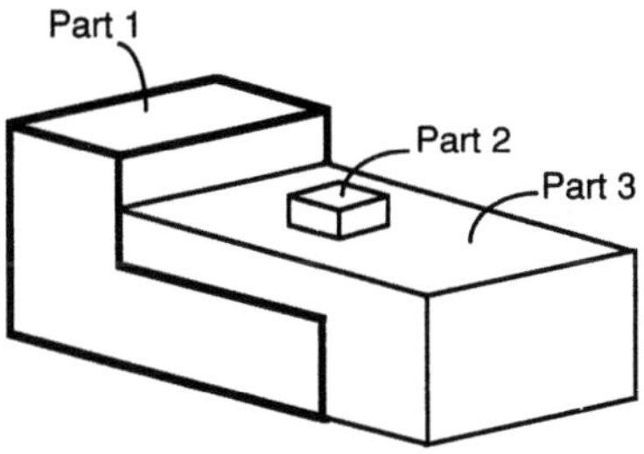

Fig. 5. A three part assembly

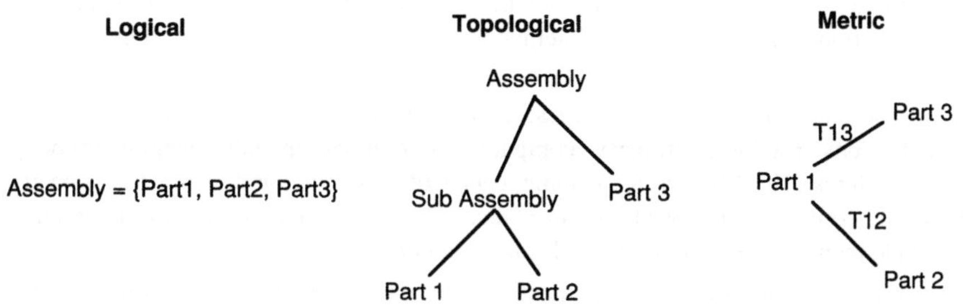

Fig. 6. Hierarchical representation for assembly in Fig. 5: Logical, Topological, Metric levels

The constraints are also represented at the logical, topological and metric levels. Consider the contact constraint shown in Fig. 7. As shown in Fig. 8, at the logical level a contact is represented by a pairs of objects in contact, at the topological level contacts are represented as contact formations. At the metric level the contacts is represented by system of algebraic equations.

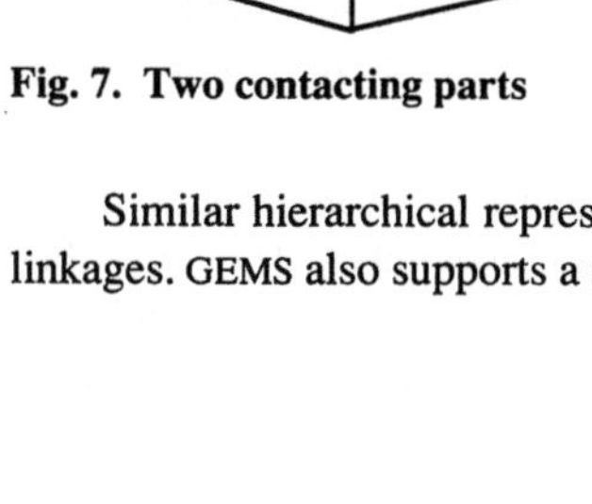

Fig. 7. Two contacting parts

Similar hierarchical representations are used to model articulated objects such as robots and linkages. GEMS also supports a simple graphics system that forms the basis for USEIT.

Logical	Topological	Metric
Part 1 contacts Part 2	F1 contact F2	$a_{11}x_1 + a_{12}x_2 + a_{13}x_3 < b_1$
		$a_{21}x_1 + a_{22}x_2 + a_{23}x_3 < b_2$
		$\vdots$

Fig. 8. Hierarchical representation of contact constraint for contact shown in Fig. 7

4. GRS: Geometric Reasoning System

As mentioned above, geometric reasoning system is a collection of geometric algorithms, geometric and motion analysis routines, and control routines. In addition GRS also has data transformation functions, and functions for interacting with the Handey robot motion planning system.

The geometric algorithms are of two types, namely property computation algorithms and geometric interactions routines. Property computations routines include integral property computation procedures [10, 18], such as, volume, center of mass, etc., and, geometric property routines, such as, convex hull, swept volumes etc. [7, 11, 17]. The geometric interaction algorithms include intersection computations, distance functions etc.

The control routines include, selection functions and data transformation routines. As we discussed in examples above, selection routines are used to select data from a set by application of constraints to members of that set. Data transformation routines are invoked when it is necessary to transform data from one representation to other. For example if a user specifies a contact between a vertex of on object and face of another object. The data transformation routines can convert this topological information to equivalent algebraic inequalities.

As we mentioned above, GEMS has special representation of contacts. The geometric reasoning system can reason about contacts from first principles to derive several very useful properties of objects in contact, such as, separation and compliant directions. Set of routines in GRS that perform such analysis and analysis of degrees of freedom of systems, static force analysis etc. are collectively know as geometric and motion analysis routines. These include routines used to analyze the kinematics of articulated objects, relative degrees of freedom among contacting objects etc. One of the crucial components in the analysis of a design or in robot task planning, is motion analysis. GARE is interfaced with a robot motion planning system Handey [12], which was developed at MIT. Handey system provides path planning and grasping planning capabilities to GRS.

Several geometrical problems lend themselves to simple algebraic formulations. The algebra engine, a subsystem of the geometric reasoning system, is a tool for solving simple algebraic problems such as solving a set of linear equations.

Geometric reasoning system also has an assembly analysis engine. GRS can generate simple assembly and disassembly sequences given the model of an assembly. The assembly/ disassembly sequences are generated from local analysis of contacts and by interacting with the path planner. Only geometric knowledge is used. More complete treatment of this problem can be found in [8, 20].

5. TDL: Task Description Language

As we mentioned above, all interactions with GARE must be done using the task description language (TDL). The TDL constitutes a collection of procedures through which a user or other systems can interact with GARE. TDL commands can be presented to GARE graphically, on command line or via function calls. Typically, graphics and command line input are given by a user, while interacting systems such as robot task planner or diagnosis system interact with GARE via function calls.

As shown in Fig. 9. There are three modes in TDL, namely, the command mode, query mode and constraint specification mode. In the command mode, the interacting system can directly command the GARE subsystems to perform some modeling or reasoning task. For example the user can issue a command to GEMS to create a model or set viewing parameter in the graphics system.

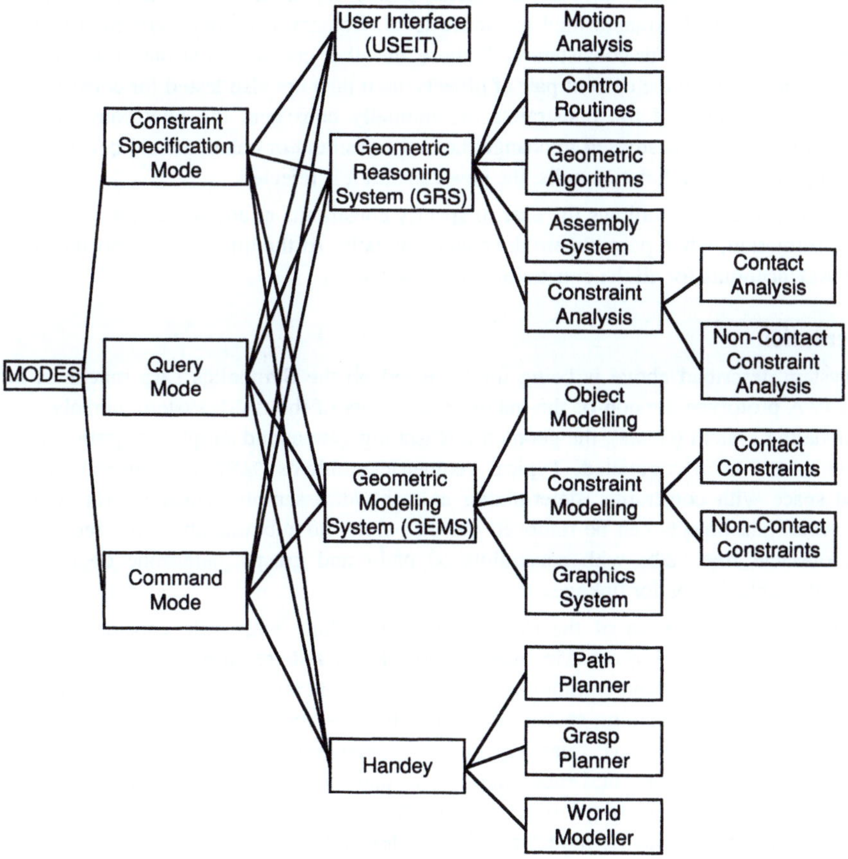

Fig. 9. Task Description Language (TDL) with three modes

All TDL interactions are processed by task description language interpretation module (TDLIM). TDLIM parses the TDL inputs and depending on whether it is a command, a query or a constraint specification, it either directly sends it to the appropriate module, or to GARE task specification translator that was discussed earlier.

There are two types of queries, passive and active. The passive queries seek information that is already available in the system and only needs to be looked up such as topology of a predefined assembly. Therefore passive queries are directly passed to the appropriate modules of GARE. The active queries require computation and they are resolved by GARE as task specifications. For instance queries to determine if an object is convex or whether a linkage system is statically stable require computation. These queries are translated into task specifications by TDLIM.

In the constraint specification mode, constraints are posted in appropriate module by TDLIM. Suppose a user specifies a contact between the vertex of one object and face of another object, the constraint is immediately transformed to metric level. Constraint is then verified with the respect to the geometric data for consistency. If there are other constraints at metric level that were predefined and apply to the current pair of objects, then they are also tested for consistency against the new constraint. If the constraints are mutually consistent the new constraint is accepted, otherwise the user is advised to change the new constraint or delete the old constraints. If the interacting system uses function calls, the new constraint is rejected.

As mentioned before TDL allows the user to specify constraints at any level of abstraction. The data transformation, when possible are done automatically. At the time of task specification translation, GARE accounts for all the constraints in the system.

6. Implementation

The GARE system described above is being implemented on the Symbolics Lisp machines in COMMON LISP. A prototype version of the major components of the GARE system, namely, the geometric modeling system (GEMS), the geometric reasoning system and simple integration with Handey system have been completed. Typical geometric problems GARE can currently solve include, find space with constraints, determining assembly/disassembly sequences for simply separable objects (i.e., objects can be removed one at a time; no sub-assemblies are required), determining obstacle free paths with constraints on paths and grasps, determine degrees of freedom and allowable ranges for linkages.

The current implementation of the GARE system was done with little consideration to computing efficiency. Typical geometric queries can take anywhere from a few seconds to several minutes depending on the complexity of the task specification. The tasks that involve interaction with the motion planning system are relatively slow because, first the path planning and grasp planning algorithms have relatively high computational complexities, secondly, the Handey system uses a different solid modeling system and therefore, considerable amount of computation time is spent in translation of data from the GEMS representation to the representation used by the geometric modeler in the Handey system.

As we described above, whenever a constraint is presented to GARE, it tries to check for it's to verify it's consistency with the existing knowledge. In case of geometric and contact constraints, this requires simultaneous solution to a system linear equalities and inequalities. At first, we implemented the algebra system such that it would solve for equations symbolically. However, this was found to be computationally very expensive. The new version of algebra engine uses fast optimization techniques to converge to solution. While it computationally more efficient, it is not guaranteed to find a solutions as the symbolic algebra system did.

The prototype GARE system has been tested for simple robot task plans and motion planning problems.

7. Conclusions

GARE is an evolving system. Current emphasis is on identifying the problems and merits of this approach. The prototype implementation has been successfully used to design and reason with simple robot task planning scenarios that included linkages and assemblies.

The future research is aimed at extending the modeling and the reasoning systems to include curved surfaces, deal with representation of and reasoning with the uncertainty in geometry. Often the spatial models are constructed not by user, but by an external sensing system, such as a vision system or a range imaging system. However, the data available from these sensing modalities can only give information about spatial occupancy. We will explore the issues of modeling and reasoning using multiple representation schemes.

Geometrical modeling is used in numerous other domains such as automated manufacturing, computer vision etc. Automated systems must be able to reason about geometrical interaction. Therefore concept of GARE should be applicable in a wide variety of domains.

References

1. A. Baer, C.M. Eastman, and M. Henrion, "Geometric Modeling: A Survey," *Computer Aided Design*, vol. 11, no. 5, pp. 253-272, 1979.
2. B. Baumgart, "Geometric Modeling for Computer Vision," Ph.D. Thesis, Stanford University, 1974.
3. B. Baumgart, "A Polyhedron Representation for Computer Vision," *Proc of AFIPS Conf.*, pp. 589-596, 1975.
4. J. Boyse and J. Gilchrist, "GMSolid: Interactive Modeling for Design and Analysis of Solids," *IEEE Computer Graphics and Applications*, vol. 2, no. 2, pp. 86-97, 1982.
5. R.S. Desai, "On Fine Motion in Mechanical Assembly in the Presence of Uncertainty," Ph.D. Thesis, The University of Michigan, Ann Arbor, 1988.
6. R. Doshi, R. Desai, R. Lam, and J. White, "Integration of Artificial Intelligence Planning and Robotic Systems with AIROBIC," *Proc. IMAC Conf on Expert Systems for Numerical Computing*, 1988.
7. I.D. Faux and M.J. Pratt, *Computation Geometry for Design and Manufacture,* John Wiley, New York, 1979.

8. L. Homem de Mello and A.C. Sanderson, ''Planning Repair Sequences Using AND/OR Graph Representation of Assembly Plans,'' *Proc. IEEE Int. Conf. Rob. Auto.*, pp. 1861-1862, 1988.

9. R. Lam, et al., ''Diagnosing Faults in Autonomous Robot Plan Execution,'' *Proc. SPIE Conf.*, 1988.

10. Y.T. Lee and A.A.G Requicha, ''Algorithms for Computing the Volume and Other Integral Properties of Solid Objects.I. Known Methods and Open Issues,'' *Communications of the ACM*, vol. 25, no. 9, pp. 635-641, 1982.

11. Y.T. Lee and A.A.G Requicha, ''Algorithms for Computing the Volume and Other Integral Properties of Solid Objects.II. A Family of Algorithms Based on Representation Conversion and Cellular Approximation,'' *Communications of the ACM*, vol. 25, no. 9, pp. 642-650, 1982.

12. T. Lozano-Perez, et al., ''Handey: A Robot System that Recognizes, Plans, and Manipulates,'' *Proc. IEEE Int. Conf. Rob. Auto.*, pp. 843-849, 1987.

13. D.R. Reddy and S. Rubin, ''Representation of Three-Dimensional Objects,'' Technical Report, CMU-CS-78-113, Computer Science Department Carnegie Mellon Univ., 1978.

14. A. Requicha and H. Voelker, ''Solid Modeling: A Historical Summary and Contemporary Assessment,'' *IEEE Computer Graphics and Applications*, vol. 2, no. 2, pp. 9-24, 1982.

15. A.A.G. Requicha, ''Mathematical Models for Rigid Solids,'' Tech Memo No. 28, Production Automation Project, University of Rochester, 1977.

16. A.A.G. Requicha, ''Representation of Solid Objects-Theory, Methods, and Systems,'' *ACM Computing Surveys*, vol. 12, no. 4, pp. 437-464, 1980.

17. M.I. Shamos and F. Preparata, *Computation Geometry,* Springer-Verlag, New York, 1985.

18. H.G. Timmer and J.M. Stern, ''Computation of Global Geometric Properties of Solid Objects,'' *Computer Aided Design*, vol. 11, no. 6, 1980.

19. H.B. Voelker and A. Requicha, ''Geometric Modeling of Physical Parts and Processes,'' *IEEE Computer*, vol. 10, no. 2, pp. 48-57, 1977.

20. J. Wolter, ''On Automatic Generation of Plans for Mechanical Assembly,'' Ph.D. Thesis, The University of Michigan, Ann Arbor, 1988.

21. F. Yamaguchi and T. Tokieda, ''Bridge Edge and Triangulation Approach to Solid Modeling,'' *Proc. Computer Graphics*, Tokyo, 1984.

22. K. Yamaguchi, T. Kunii, K. Fujimura, and H. Toriya, ''Octree Related Data Structures and Algorithms,'' *IEEE Computer Graphics and Applications*, vol. 4, no. 1, pp. 53-59, 1984.

13

Constraint Satisfaction Techniques for Spatial Planning

Can A. Baykan and Mark S. Fox

Center for Integrated Manufacturing Decision Systems
The Robotics Institute
Carnegie Mellon University
Pittsburgh, PA. 15213, USA

Abstract: WRIGHT is a CAD system for designing two dimensional layouts consisting of rectangles. This problem arises in space planning, i.e. the design of floorplans, arrangement of equipment in rooms, and site planning. Space planning is a search process characterized by very large search spaces. Constraint directed search provides a basic problem solving methodology for intelligent CAD by providing a formal method for representing domain knowledge uniformly as constraints, and by using constraints for efficient search. Each configuration is represented as a Constraint Satisfaction Problem (CSP), and constraint propagation restricts the domains of variables as information becomes available during search. Constraints reduce search complexity by choosing the most constrained decision. The measures that identify opportunistic decisions are variable tightness, constraint reliance and constraint tightness. Least commitment representations (value ranges) delay decisions until enough information becomes available. Abstractions reduce complexity by allowing abstract constraints to prune away entire design subsets.
Keywords: Space Planning, Search, Constraint Satisfaction, Least-commitment, Abstraction

1. Introduction

Intelligent CAD requires a fundamental problem solving methodology that can incorporate arbitrary amounts of knowledge in a principled manner. Constraint-directed search provides a formal method for representing expertise uniformly as constraints. From constraints an understanding of the structure of the problem (search) space, that leads to more efficient search, can be derived [1, 7, 8, 10]. Thus constraint-directed search addresses the needs of intelligent CAD by enabling the representation of knowledge from diverse sources and enabling the selection of efficient search strategies based on an understanding of search space structure.[1]

[1] This differs from encapsulating expertise in the form of rules in that rules do not provide for an understanding of problem space structure, but simply identify situations of applicability without any guarantee that the search being performed is efficient.

This basic problem solving methodology provides the framework within which the following issues can be addressed:

- knowledge representation,
- acquisition and maintenance of design expertise,
- user interface for graphical specification of constraints,
- user interface for interactive design.

WRIGHT is an intelligent interactive *space* planning system for generating two dimensional layouts consisting of rectangular shapes using constraint-directed opportunistic search. Space planning deals with the design of two dimensional layouts, such as floor plans, the arrangement of equipment in rooms and site planning. In space planning, topological relations and shape, dimension, distance and other functions of spatial arrangement are a principal concern. Almost all aspects of design have spatial implications, and influence space planning decisions.

2. Problem

WRIGHT deals with the generation of two-dimensional layouts consisting of configurations of rectangles. Inputs for generating a layout are:

- An existing layout which may be an empty space, and dimensions of which may be specified as ranges, as seen in Fig. 1.

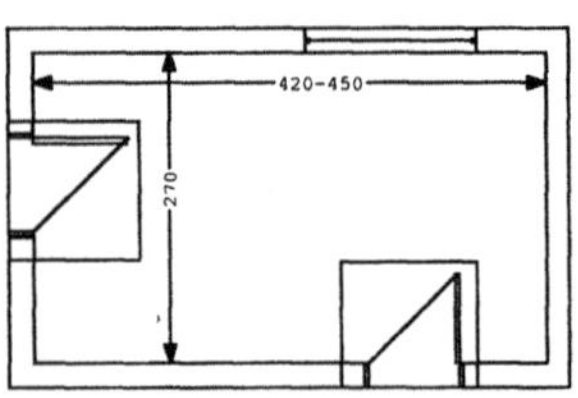

Fig. 1. Plan of kitchen showing existing layout

- Design units to be located and/or dimensioned, as seen in Fig. 2.

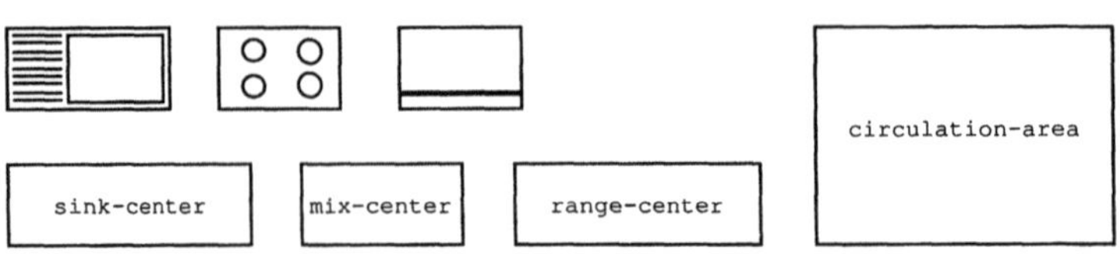

Fig. 2. Design units to be dimensioned and located in kitchen

- Knowledge about the design domain in the form of a class hierarchy of prototype design units and constraints on them, as seen in Fig. 3.

The output of WRIGHT is a set of optimal layouts that are significantly different from each other, as seen in Fig. 4.

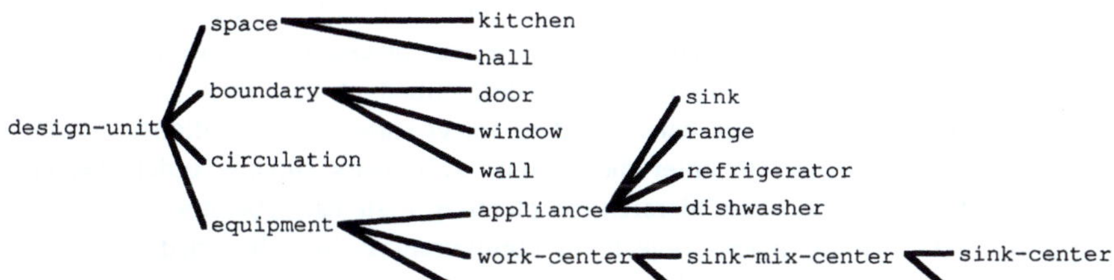

- Equipment inside kitchen
- Appliance non-overlap appliance
- Appliance completely-next-to circulation
- Work-center next-to circulation
- Range-center next-to sink-mix-center
- Sink inside sink-center
- Sink next-to window
- Sink length ≥ 120 cm.

Fig. 3. Class hierarchy of kitchen design units and some kitchen constraints

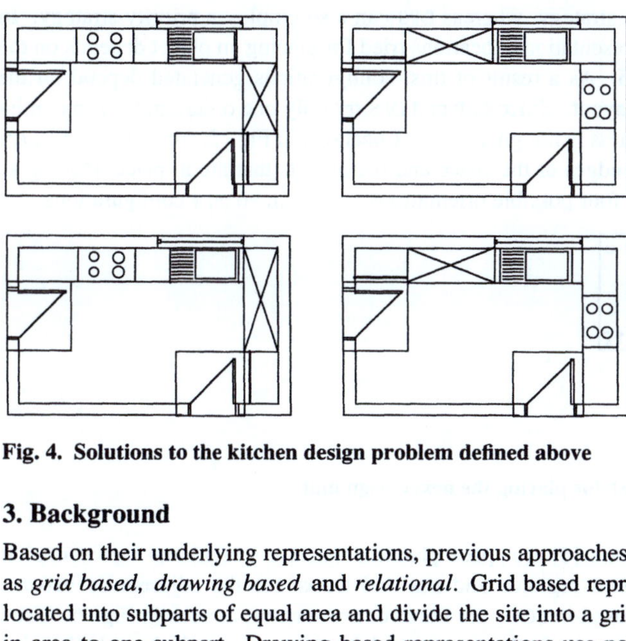

Fig. 4. Solutions to the kitchen design problem defined above

3. Background

Based on their underlying representations, previous approaches to spatial layout can be classified as *grid based, drawing based* and *relational.* Grid based representations partition objects to be located into subparts of equal area and divide the site into a grid of cells where each cell is equal in area to one subpart. Drawing based representations use polygons of fixed size and shape to represent objects. A polygon is represented as a set of sides, and a side as a set of points. Relational representations use adjacency or incidence between points, between lines and regions, or between regions to model layouts.

In space planning, search operates by selecting a spatial element(s) and an operator, and generates a new configuration by applying the operator to the element(s) in some state. Structuring the elements of search gives rise to different strategies. Starting search with an empty initial configuration results in a *build-up strategy*. There are two basic variations in a build-up strategy: *organize by element* and *priority* solution methods [3]. In an organize by element strategy, the next object to enter the layout is selected, placed at alternative locations and tested. All relevant attributes of the object are determined at the time it enters the configuration, and all applicable tests are carried out to select satisfactory locations. Search continues by selecting a new object to enter the design. A priority strategy orders search operators as in ABSTRIPS [16] and other hierarchical planning systems. Operators with high priority are applied first, creating macro objects or configurations in unbounded space by determining the important attributes first. In an *improvement strategy,* search starts from a configuration which contains all the elements. Changes are made in response to failing constraints or in order to improve the score of an objective function.

Quadratic assignment formulation [12] uses a grid based representation. This representation can not deal with variable sizes, and makes it very hard to deal with issues of shape and alignment. Both build up and improvement strategies are used with this approach.

DPS [15] and GSP [3] use drawing based representations. GSP objects must be rectangles and DPS objects can be arbitrary polygons. In both systems dimensions of objects must be fixed. GSP uses an organize by element strategy, whereas DPS can also employ a priority strategy. In systems using drawing based representations, locations tried for placing an object depends on the existing layout, as seen in Fig. 5. As a result of this, configurations generated depend on the order in which objects enter the layout. Since GSP and DPS try only one ordering, they may miss possible solutions. Their correctness is not guaranteed. Locations in Fig. 5, from ([3]; p.57), are defined by lines projected by the edges of the space and the objects that are in place. Placing an object at every location above, in four possible orientations, results in 96 new configurations.)

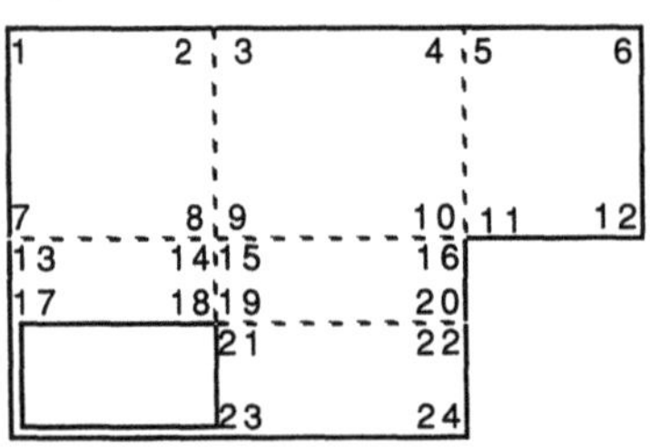

Fig. 5. Locations considered by GSP for placing the next design unit

GRAMPA [9], DIS [4] and LOOS [6] are space planners that use a relational representation. GRAMPA uses adjacencies between regions, and DIS and LOOS use the adjacencies between regions and lines. The relations used in LOOS are north-of and east-of, as seen in Fig. 6. Using these relations it is not possible to specify that rectangle 1 is next-to rectangle 2 – whether they are adjacent or not depends on the dimensions. The same is true of specifying alignment. Such relational considerations are treated as dimensional issues.

Relational layout programs systematically generate all distinct configurations defined by their representation. They use a two step process that deals with relational and dimensional aspects separately, thus are not opportunistic in the use of constraints. The relations that are

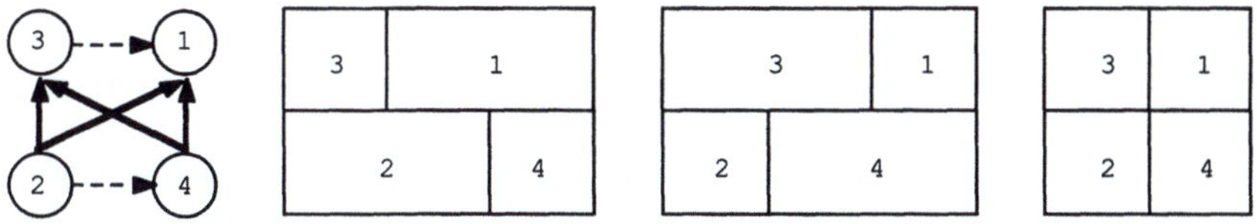

Fig. 6. An orthogonal structure and possible configurations represented by it

used for generating solutions are not based on the requirements of the problem, but on a restricted set of relations defined in each system.

WRIGHT uses relations between lines for expressing topology. It is possible to create new spatial relations by defining them in terms of relations between lines. The strategy employed in WRIGHT is a priority strategy, where search operators determine only the attributes specified by the selected constraint. The topology and dimensions of a configuration can be decided in any order due to the CSP formulation used.

4. Insights and Approach

Design is the process of constructing a description of an artifact that satisfies a functional specification, meets explicit or implicit performance criteria, is realizable and satisfies restrictions on the design process itself [13]. The artifact is initially defined by its desired properties, and it is natural to express them in terms of constraints. In WRIGHT, any knowledge that defines or restricts the domain is expressed as a constraint.

Space planing is a search process [2] characterized by very large search spaces. Constraints play a major role in reducing search complexity, by opportunistically choosing the most constrained decision to make at each step [7]. Constraint propagation restricts the set of alternatives by eliminating values that are inconsistent with the current decision. WRIGHT uses constraints to select an efficient search path, and uses constraint propagation to ensure that values of all variables are consistent.

Design representations should tolerate ambiguity and incomplete specification. They are different from representations of existing situations. Least commitment [16] representations delay decisions about uncertain aspects, while making it possible to reason about the certain aspects of designs. Least commitment is achieved by using value ranges, abstract constraints, and generating only the attributes specified in constraints.

Abstractions can further reduce complexity by allowing abstract constraints to prune entire design subsets [14]. Abstract constraints bound the solution space, and objects at different levels of aggregation simplify search in WRIGHT. Decisions are not forced by the representation used. When a commitment is made, it is always for satisfying a constraint. Attributes of design not specified by the selected constraint are deferred.

The goal of this research is to identify and elaborate the semantics of constraints to express knowledge of the problem domain, to identify the knowledge that enables efficient search decisions, and to find the limitations, applicability and performance of this approach with respect to type and structure of constraints.

5. Representation

The elements of WRIGHT are design units (objects), spatial relations and constraints. Knowledge is represented by a class hierarchy of prototype design units, and constraints specifying desired relations between design units or restrictions on their attributes such as length or orientation. *Abstraction by aggregation* combines design units into larger objects, such as combining a set of rooms into a house; or a sink, dishwasher, and counter-top area into a sink-center. WRIGHT can handle problems involving design-units at different levels of aggregation.

5.1. Objects

There are two types of objects used in WRIGHT for representing layouts: *numerical variables* and *structured objects.* Structured objects contain other structured objects or numerical variables. Numerical variables are *interval* or *discrete.* Design units and rectangles are structured objects. Lines, dimensions, areas, and distances are interval variables. Orientations are discrete variables.

Design unit is the building block of configurations. A design unit is a structured object made of two components: a rectangle and an orientation, as seen in Fig. 7. Design units have fronts, backs and sides. Orientation of a design unit specifies which way its front is facing. The counterclockwise angle between the front of the design unit and the y-axis is its orientation. A rectangle is a structured object consisting of four lines, length, width, and area.

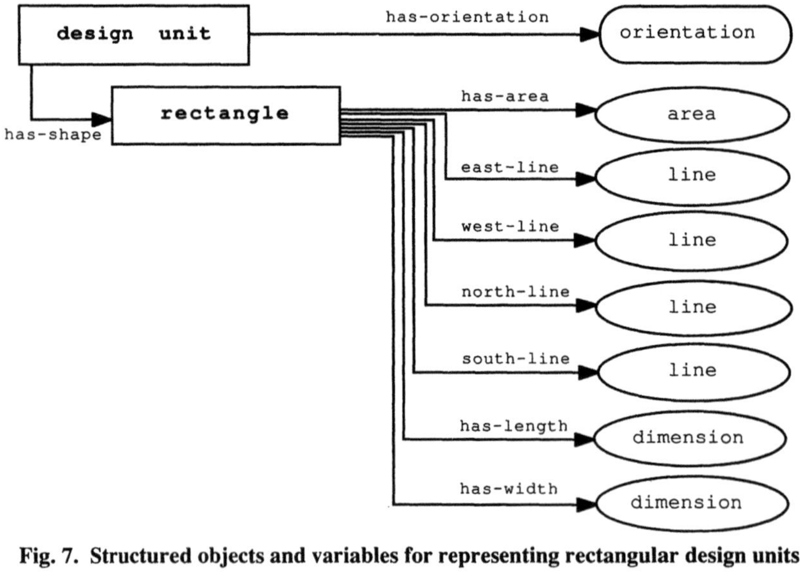

Fig. 7. Structured objects and variables for representing rectangular design units

Discrete variables take a single value selected from a set of allowable values. Orientation is discrete variable, and the set of allowable values for it are: { 0, 90, 180, 270 }. Interval variables have a range of values defined by a minimum and a maximum. For vertical lines, minimum and maximum values bound the x-coordinate of the line. For horizontal lines, they bound the y-coordinate.

5.2. Relations

Spatial relations express the types of configurations that are of interest in space planning problems. Spatial overlap and adjacency relations are topological relations where orientations or relative locations with respect to global or object centered coordinates are not considered.

- **Spatial overlap:** inside, contains, overlaps, non-overlapping, one-dimensional-overlap.
- **Adjacency:** next-to, completely-next-to, covers.

Location and alignment relations listed below are defined with respect to the global directions: north, south, east, and west. Orientations of the design units are not considered. For example, align-north specifies that the north sides of two design units are to be aligned, regardless of their orientations.

- **Location:** north-of, south-of, east-of, west-of.
- **Alignment:** align-north, align-south, align-east, align-west.

The following spatial relations are defined with respect to object centered coordinates, and depend on the orientation of one or both design units.

- **Orientation**: parallel-to, perpendicular-to, opposite.
- **Relative location**: in-front, at-back, at-left, at-right.
- **Relative alignment**: align-front, align-back, align-left, align-right.
- **Relative distance**: front-distance, back-distance, left-distance, right-distance.

Orientation relations specify relative orientations of two design units. If the orientations are equal, the design units are parallel. If the orientations differ by 90°, the objects are perpendicular. If the orientations differ by 180°, the objects are opposite. Relative location relations specify the location of design unit 2 with respect to design unit 1. Relative alignment relations specify that sides of two design units are collinear. Relative distance relations specify the distance from the specified side of design unit 1 to the closest side of design unit 2.

Spatial relations between design units are defined in terms of and/or combinations of algebraic relations between their lines. The algebraic relations are: =, >, ≥, +, and ×. The statement vertical-line1 > vertical-line2 means that vertical-line1 is to the east of vertical-line2, since the value assigned to a vertical line specifies its x-coordinate.

There is a grammar for defining spatial relations, which maps the relation into disjunctive and conjunctive combinations of algebraic relations between numerical variables or constants. As a result of observing the problem solving behaviour of WRIGHT on kitchens, we have identified new spatial relations such as: completely-overlapping, one-dimensional-overlap, at-side in addition to the ones we have started with and included them in the system.

5.3. Representing layouts

A layout is represented by a set of variables, their values, and algebraic relations that should always be satisfied. This defines a Constraint Satisfaction Problem (CSP). A CSP has a set of variables, each with an associated domain, and a set of constraining relations, each involving a subset of the variables. The variables in the CSP representation of a layout are the lines, dimensions, areas and orientations of the design unit instances, plus other variables that indicate distances or amounts of overlap. Types of relations are: =, >, ≥, +, and ×. The first three relations are either unary relations between a variable and a constant or binary relations between two variables. The last two are ternary relations between three variables.

The constraining relations in a CSP are defined by rectangles on their component variables, and by topological and geometrical relations between design units. A rectangle defines constraints between its lines, dimensions and area, as seen in Fig. 8.

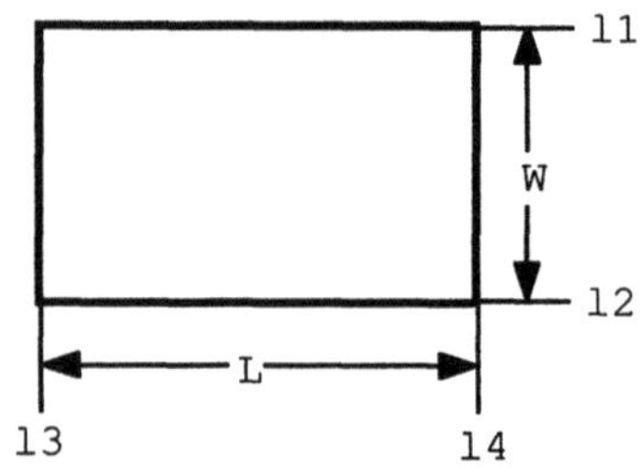

l1 + W = l2

l1 < l2

l3 + L = l4

l3 < l4

L * W = Area

Fig. 8. Constraining relations defined by a rectangle

Setting up relations between design units results in algebraic relations between their components. In Fig. 9, the sink defined by lines ln1, ln2, ln3, ln4 is placed next-to and south of the window defined by lines ln5, ln6, ln7, ln8. This configuration should satisfy the constraint that sink is next-to the window for ≥ 50 cm. The algebraic constraints seen in Fig. 9 are added to the CSP as a result. Variables v1 and v2 are created for expressing the adjacent distance between sink and window.

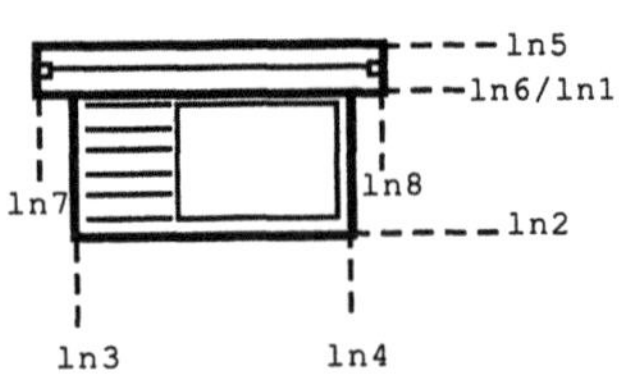

ln1 = ln6

ln7 + v1 = ln4

ln3 + v2 = ln8

v1 ≥ 50

v2 ≥ 50

Fig. 9. Constraining relations defined by a configuration

5.4. Constraints

Any knowledge that defines or restricts the domain is a constraint. Constraints indicate restrictions or desired relations. In WRIGHT constraints are of the form:

- <object> <relation> <object>, or
- <variable> <algebraic relation> <number>.

Constraints expressing domain knowledge are posted to prototype design units, as seen in Fig. 3. These are called *domain constraints*. In addition to specifying a relation or a restriction on some value, a constraint expresses knowledge about how it is to be used during search. This knowledge is expressed by *similarity* and *relaxation* relations between constraints and *combining-instances, importance,* and *utility* values.

When it is posted, each constraint is assigned an *importance* value specifying the importance of satisfying that constraint in a solution. All constraints are binary, so they are either satisfied or contradicted. Solutions are rated by subtracting the importance values of failing constraints. *Relaxations* of constraints are specified by other constraint expressions that specify alternative relations, alternative design units, or looser bounds on numerical variables. Relaxations are tried when a constraint can not be satisfied in its original form. States where a constraint is relaxed are assigned lower ratings as specified by the *utility* of the relaxation.

Design knowledge is expressed in terms of required spatial relations in WRIGHT. Consider the relationship of the sink to windows: "The average housekeeper spends nearly 1 and 1/4 hours at the sink each day so there is a good case for putting the sink at a window for good light and view." ([11]; p.72) One way of satisfying the requirements is placing the back of the sink completely next to the window, which is expressed by the following constraints:

- *Sink completely-next-to window*
- *Sink at-back window*

When it is not possible to put the sink completely next to the window, placing it in front of and perpendicular to the window will allow direct light and a view of outside. The sink must also be close enough to the window. The following constraints express this case:

- *Sink distance window, max=120 cm.*
- *Sink one-dimensional-overlap window, min=30 cm.*
- *Sink perpendicular-to window.*

Distance is measured between closest points. One dimensional overlap means overlap in either the vertical direction or horizontal direction. The second set of constraints are a relaxation of the first set, and have lower utility values.

When there is more than one window, the constraints between sink and window should be satisfied for one window only. Whether the constraint should apply to all windows or just one window is indicated by *combining-instances* value attached to the constraint or to the relation specified in the constraint. There are defaults for relations, i.e. for non-overlap the default value for combining-instances is *and* whereas for most other relations the default is *or*.

Similarity is a relation between two constraints which restricts the way the constraints apply in case there is more than one instance. In the example above, sink should be completely next to and have at its back the same window. This is expressed by posting a *similarity* condition between the two constraints.

5.5. Constraint graph

The constraint graph is an and/or network that refines design knowledge represented by constraints on prototype design units into a design specification represented as combinations of algebraic constraints on the components of the design unit instances.

A constraint graph consists of nodes and links as seen in Fig. 10. (The constraint *sink next-to window* is mapped to constraints on the components of sink1 and window.1.) It is an abbreviated graph, where a constraint in the CSP and the variables it connects, such as: line1=line2, are represented by a single node. Nodes are of two types: and-nodes, or-nodes. And-nodes are expressed by connecting the links leaving that node by an arc. The links in the constraint graph indicate *reliance* between constraints.

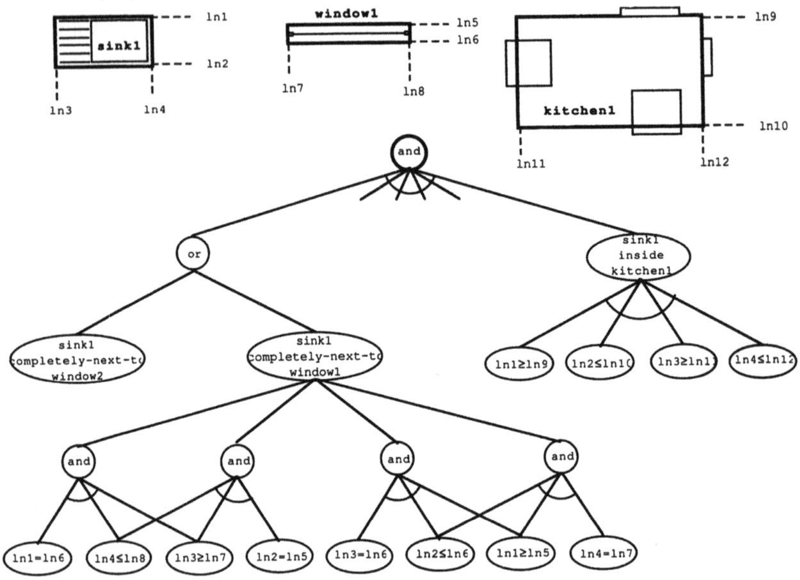

Fig. 10. Partial constraint graph

The constraint graph specifies alternative ways of satisfying a constraint. Prototype design units that have more than one instance and spatial relations that can be satisfied in different ways introduce disjuncts to the constraint graph. The top level of the graph is in *conjoint normal form.*

When there are conditions which hold true in all the alternatives, they can be used to bound solutions without committing to a specific alternative. These are called abstract constraints. Abstract constraints exist for adjacency and distance relations, and for dimensional constraints. Fig. 11 shows the abstract constraints for the constraint graph seen in Fig. 10. The configuration in the figure shows the location defined by the abstract constraints for the sink when the location of the window is fixed.

6. Search Architecture

The search architecture used in WRIGHT operates by selecting a node connected to the top level of the constraint graph and satisfying it in all possible ways. Operators are associated with constraints as in means-ends analysis. The variables indicated by the constraint are assigned new values, constraints are propagated, and any constraints that have only one way of being satisfied are also satisfied. Dealing with variables that are not affected is deferred until later. This is a priority strategy where *texture measures* select the constraint to be satisfied. The

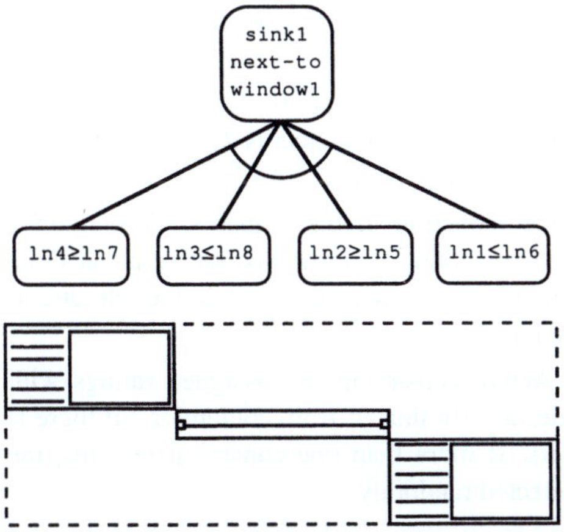

Fig. 11. Abstract constraints for *sink1 next-to window1* and resulting bounds on location of sink

constraint graph provides a mesh over the problem space and is used for generation and testing. Texture measures provide an approximation of problem space topology that allows search to be focused.

6.1. Focus of attention algorithm

Given an initial configuration and design units to locate and dimension, search starts by compiling the constraint graph. Every search state contains a CSP representation of the layout, and a constraint graph with values based on the CSP. New layouts are generated by selecting some constraints from the constraint graph, and adding them to the CSP. Search is controlled by the focus of attention algorithm, which operates as follows:

1. The state to continue search from is selected using a *best first* strategy.
2. In a selected state, one of the constraint nodes connected to the root of the constraint graph is selected, based on the problem space textures described below.
3. A new state is generated for satisfying each disjunct of the selected constraint. In each new state, constraints that contain no disjuncts are satisfied. Satisfying a constraint can determine values of disjuncts in other constraints, therefore this step loops until quiescence.
4. When the status of all constraints in the conjoint normal top level of the constraint graph are determined, that state is a solution. States are rated by subtracting importances of violated constraints at top level of constraint graph.

6.2. Texture measures

WRIGHT uses three texture measures for selecting constraints. The first measure is from a variable perspective, and the last two are from a constraint perspective. The texture measures are:

- **Variable Tightness:** is approximated by the number of remaining conjunctive constraints on each design unit instance.
- **Constraint Reliance:** is approximated by *1/number of disjuncts.* Constraints with fewer disjunctive cases remaining are selected. If the number of disjuncts is 1, the constraint is satisfied without needing to generate a new state for each disjunct.
- **Constraint Tightness:** is approximated by the reduction in the domains of continuous variables involved, as a result of satisfying the constraint. Constraints that result in large reductions are favoured. Types of algebraic relations that will be added to the CSP due to each of the competing constraints are also taken into account.

Texture measures are applied lexicographically. Active constraints are assigned ratings with respect to a metric, and constraints with lower values are eliminated from contention. If there is a single constraint with the best measure, it is used. If more than one constraint remains, the next texture measure is applied, or a constraint is selected randomly.

6.3. Search operations

Each state contains a different layout and a different constraint graph, as seen in Fig. 12. When generating a state, the layout is changed by operators that satisfy a constraint. The operators carry out one or both of these actions:

- add new relations between objects,
- assign minimum or maximum values to line locations, dimensions, areas, and distances.

As a result of adding new relations or values, constraints are propagated to ensure that the values are globally consistent. Constraint propagation removes inconsistent values at continuous variables by increasing the minimums and decreasing the maximums. The constraint graph is changed by marking the nodes as *satisfied* or *contradicted.* These labels propagate up through the constraint graph according to rules of propagation defined for and-nodes and or-nodes.

Properties of the search architecture and representation of WRIGHT are:

- Search is *monotonic.* States are generated by adding new relations to the CSP, Therefore a requirement that is satisfied can not be violated later.
- Disjuncts specified in the constraint graph are *mutually exclusive.* Therefore, it is not possible to get duplicate solutions.
- Search efficiency depends on the order constraints are satisfied. Satisfying a set of constraints at the leaves of the goal tree in any order leads to the same solution.

Each solution satisfies a different subset of the constraints at the leaves of the constraint graph. Fig. 4 shows the set of solutions for the kitchen seen in Fig. 1. Mix-center is the rectangle with diagonals. In the two solutions on the left, the order of design units clockwise starting from top left is: range-center, sink-center, mix-center and refrigerator. The configuration of sink-center and mix-center at the top left corner of the kitchen is different between the solutions on the top left bottom left in Fig. 4. In the two solutions on right, the order of design units clockwise starting from top left is: refrigerator, mix-center, sink-center and range-center. The difference between the top and bottom solutions is again the configuration at the corner.

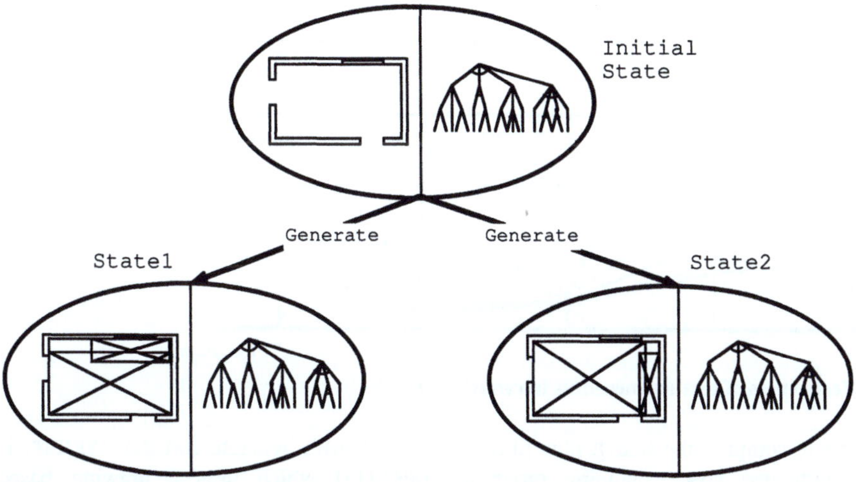

Fig. 12. Search states

WRIGHT uses constraints to define significant differences between alternatives. In this case significant differences are defined by different adjacencies and by different ways of placing the objects adjacent to each other at corners.

The attributes of the layout that are not constrained are treated as unimportant. This permits solutions at a higher level of abstraction than in other space planning systems, while enabling exact determination of relevant aspects. Unless the constraints force the assignment of a unique value, a variable has a range of values even in a solution. It is possible to refine a solution further by assigning unique values to all variables.

6.4. Performance

Texture measures reduce search. As a result, WRIGHT looks at a smaller number of states. Fig. 13 shows the number of search states required for finding all solutions to five kitchen layout problems, under different combinations of texture measures. The combinations tested are:

- *method 0:* select a constraint at random,
- *method 1:* variable tightness,
- *method 2:* constraint reliance,
- *method 3:* variable tightness and constraint reliance,
- *method 4:* variable tightness, constraint reliance and constraint tightness.

When a combination of measures is used, they are applied in the order: variable tightness, constrain reliance, constraint tightness. Each measure eliminates some constraints from consideration. If more than one constraint remains after applying the texture measure(s), specified by the method, a constraint is selected at random. The number of states given for each problem-method combination is the average of three runs. In the second problem, method 4 reduces search by more than 80% compared to method 0, and in the third problem by 35%.

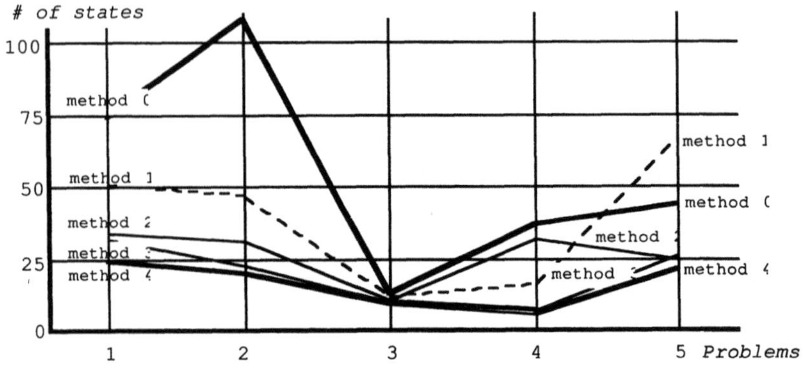

Fig. 13. Effectiveness of texture measures in reducing search

In order to compare the search plus filtering approach with generate and test, WRIGHT is compared with two space planning programs: DPS [15] which uses a drawing based representation, and LOOS [5] which uses a relational representation.[2] The problem used in the comparison is arranging six fixed size blocks in a box such that no blocks overlap. Due to the simplicity of the problem, exactly the same set of constraints can be used by all three programs. The programs are compared in terms of the number of states and search plies[3] generated when finding the first solution and when finding all 24 solutions, seen in Table 1.

	First Solution		All 24 Solutions	
WRIGHT	5 plies	14 states	5-6 plies	119 states
LOOS	6 plies	68 states	6 plies	232 states
DPS	6 plies	72 states	*(not available)*	

Table 1. Comparison of WRIGHT, DPS and LOOS in terms of search efficiency

In DPS and LOOS, the number of search plies is always equal to the number of objects to be located, as a result of the organize by design unit strategy. WRIGHT's performance in terms of number of search plies and number of search states depends on number and strength of available constraints and their interactions. Although the constraints in this problem are not as varied as in kitchen layout, WRIGHT performs better than DPS and LOOS. WRIGHT looks at a smaller number of search states by selecting decisions with fewer alternatives, and by eliminating inferior alternatives earlier.

Another area of comparison is how search behaviour changes when problems are under or over constrained. In an underconstrained problem, DPS and LOOS find the first solution faster, but there will be a large number of solutions. WRIGHT also finds the first solution faster, and will avoid generating a large number of solutions by having solutions at a higher level of

[2] See §3 **Background** for a discussion of these programs.

[3] The number of search plies is the number of intermediate states on a path from the initial state to a solution state.

abstraction. In an overconstrained problem, DPS will not be able find any solutions because it rejects a solution that fails any constraint. For LOOS, overconstrained problems pose the same difficulty as underconstrained ones: too many states with equivalent scores. Finding the first solution will take much longer too. Overconstrained problems will cause WRIGHT to search longer before finding the first solution. When all constraints can be satisfied, solutions are defined by alternative ways of satisfying all constraints. When all constraints can not be satisfied, combinations of constraints that result in equal ratings need to be tried. By defining explicit relaxations for some domain constraints in its knowledge base, WRIGHT avoids searching a large number of constraint combinations.

7. User Interface

WRIGHT's model of the design process and the division of labour between system and user is that the user inputs the problem, identifies, modifies or relaxes constraints, and the system finds solutions. The user may also make search control decisions.

The user interface provides a menu-driven, graphical means for designing, posting constraints and carrying out search. The user interface is designed to facilitate:

- examining partial solutions,
- adding missing constraints,
- relaxing overly restrictive constraints,
- selecting which partial solution to pursue next,
- observing search decisions and reasons for it.

A change such as adding or removing a constraint and bounding the location of a design unit takes place in the current search state and only affects states that are generated from it.

Fig. 15 shows three windows: for graphics, for the search tree, and for text. There is a command menu in the top left corner. When the user selects a command by clicking on it with the mouse, menus pop up for selecting the parameters of the command. A pop-up menu for selecting design unit instances is seen in Fig. 15, superimposed on the plan.

The **select-state** command brings the search tree window in the bottom right corner of the screen to the top, and enables the user to select a state by pointing and clicking with the mouse. The configuration in the selected state is displayed in the graphics window, and following operations take place in that state. The next two commands, **interactive-search** and **exhaustive-search**, are for generating a set of new states from the current state and for carrying out exhaustive search.

There are commands to **create**, **size**, **locate**, and **orient** design units. The designer can define a rectangle by clicking at its top left and bottom right corners in the graphics window using the mouse. Rectangles are used to input minimum size, maximum size and location of objects. It is possible to think of a rectangle as a constraint, because it indicates bounds. During interactive sizing and locating operations WRIGHT will not allow the user to violate existing bounds on a design unit. For relaxing bounds, one needs to move up in the search tree to a state where those variables have looser bounds. Constraints specifying relations between design units at any level of the class hierarchy, including particular instances of design units, are posted by

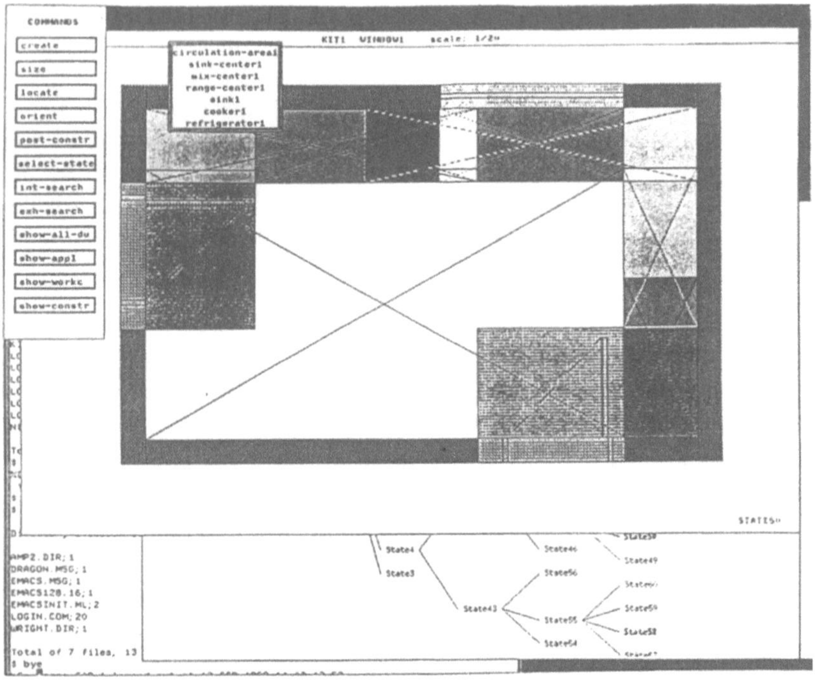

Fig. 15. WRIGHT's user interface

first selecting **post-constraint** from the command-menu, and then selecting a design unit, a relation and another design unit from pop-up menus.

The rest of the commands are for displaying information: displaying all design units – as seen in Fig. 15, displaying only the appliances, and displaying only the work centers. **Show-constraints** command displays the constraints that have not been satisfied yet, in the text window at the bottom left hand corner of the screen. It is also possible to inspect the database and call functions, as the text window is the lisp listener.

8. Conclusion

WRIGHT has been tested on kitchen design, house layout and blocks problems. The representation scheme used in WRIGHT is more flexible than other space planning systems in two respects:

- It allows design units at different levels of aggregation.
- It enables declaratively defining new relations.

If, in some domain of application, we identify new relations that express the conditions we are interested in, they can be defined as combinations of =, >, ≥, +, and × relations between lines and can be used in constraints.

The opportunistic constraint directed search approach of WRIGHT leads to significant increases in performance relative to other spatial planning techniques. This is due to two factors. First, WRIGHT uses knowledge of constraints to develop a mesh over the problem (search) space. Propagation of restrictions occurs within the mesh resulting in a reduction in size of the problem space. Second, WRIGHT opportunistically selects variables to constrain once propagation ends. Knowledge of problem space 'texture' is used to identify the appropriate variable. Texture measures used are: *variable tightness,* which chooses a design unit having a large number of constraints; *constraint reliance* which selects a constraint having fewer disjunctive alternatives, and *constraint tightness* which selects a constraint that reduces the domains of its variables most [8].

In summary, the philosophy behind this approach is to use constraints to understand the structure of the search space to make search efficient. WRIGHT generates fewer alternatives as a result of selecting decisions opportunistically, avoiding premature commitment, and eliminating inferior alternatives earlier. Constraints guide generation of significantly different alternatives. Insignificant aspects of the design do not cause alternatives to be generated while relevant differences are explored in all possible ways.

References

1. R. Dechter and J. Pearl, "Network-Based Heuristics for Constraint-Satisfaction Problems," *AI*, vol. 34, no. 1, pp. 1-38, 1988.
2. C.M. Eastman, "On the Analysis of Intuitive Design Processes," in *Emerging Methods in Environmental Design and Planning*, ed. G.T. Moore, MIT Press, Cambridge, MA, 1970.
3. C.M. Eastman, "Automated Space Planning," *AI*, vol. 4, pp. 41-64, 1973.
4. U. Flemming, "Wall Representations of Rectangular Dissections and their Use in Automated Space Allocation," *Environment and Planning B 5*, pp. 215-232, 1978.
5. U. Flemming, "On the Representation and Generation of Loosely Packed Arrangements of Rectangles," Tech. Rept. DRC-48-05-85, Carnegie-Mellon University Design research Center, 1985.
6. U. Flemming, M.D. Rychener, R.F. Coyne, and T. Glavin, "A Generative Expert System for the Design of Building Layouts," Center for Art and Technology, CMU, Pittsburgh PA, 1986.
7. M.S. Fox, "Observations on the Role of Constraints in Problem Solving," *Proceedings Sixth Canadian Conference on Artificial Intelligence*, pp. 172-187, 1986.
8. M.S. Fox, N. Sadeh, and C. Baykan, "Constrained Heuristic Search," *Proceedings of IJCAI-11*, pp. 309-315, 1989.
9. J. Grason, "Methods for the Computer-Implemented Solution of a Class of Floor Plan Design Problems," Ph.D. Th., Carnegie-Mellon University, 1970.
10. R.M. Haralick and G.L. Elliott, "Increasing Tree Search Efficiency for Constraint Satisfaction Problems," *AI*, vol. 14, pp. 263-313, 1980.
11. Architects Journal, "Domestic Kitchen Design: Convential Planning," *Architecs Journal*, pp. 71-78, 1984.

12. J.C. Koopmans and M.J. Beckmann, ''Assignment Problems and the Location of Economic Activities,'' *Econometrica*, vol. 25, pp. 53-76, 1957.

13. J. Mostow, ''Towards Better Models of the design Process,'' *AI Magazine*, vol. 6, no. 1, pp. 44-57, 1985.

14. A. Newell, J.C. Shaw, and H.A. Simon, ''The Processes of Creative Thinking,'' in *Contemporary Approaches to Creative Thinking*, ed. H.E. Gruber, G. Terrel, and J. Wertheimer, Atherton, 1962.

15. C. Pfeffercorn, ''Computer Design of Equipment Layouts Using the Design Problem Solver,'' Ph.D. Th., Carnegie-Mellon University, 1971.

16. E.D. Sacerdoti, ''Planning in a Hierarchy of Abstraction Spaces,'' *AI*, vol. 5, pp. 115-135, 1974.

14

A Geometric Constraint Management System in Oar

Farhad Arbab, and Bin Wang

Computer Science Department
SAL 200, MC 0782
University of Southern California
Los Angeles, CA 90089, USA

Abstract: This paper presents a constraint management system to experiment constraint-based design for mechanical engineering applications. The design is object-oriented and constraint-oriented. A design process can be viewed as incrementally creating design objects and imposing constraints on them. The system provides a sketch model and a constraint model which are loosely-coupled so that a designer can evolve a design by evolving both models separately. This system architecture is more appropriate for interactive design than existing systems. We use Operational Transformation Planning (OTP) as an approach to satisfy a network of geometric constraints [2]. The approach is different primarily in its use of geometric reasoning for satisfaction planning. An OTP system automatically generates a sequence of operations to satisfy a network of constraints, based on its understanding of the higher-level semantics of constraints and the geometric implications of the operations. Our system is implemented using Oar, an object-oriented programming environment with symbolic reasoning capabilities.
Keywords: Constraints, Constraint Satisfaction, Operational Transformation Planning.

1. Introduction

The increasing demands for design automation present the challenging goal of adding 'intelligence' to CAD systems, especially in the domain of computer-aided mechanical design where satisfying geometric constraints is a major concern. We expect an intelligent CAD system to be sufficiently knowledgeable in the domain of geometry to 'understand' the implications of geometric constraints, and to find appropriate means to satisfy a set of constraints as a design evolves.

Constraint-based design systems have been developed in the past, some, specifically to deal with geometric constraints. Among them, the systems based on algebraic interpretation of geometric constraints seem dominant, e.g., the work of Hillyard [8] and variational geometry [13-15]. The algebraic approach translates dimensional constraints into a system of algebraic equations (non-linear equations generally) and finds an embedding of the geometry by solving the system of equations numerically. The algebraic approach sacrifices the more abstract semantics of geometric constraints for numerical computation, leading to problems including (i) large amount of computation is required when the number of equations and variables grows; (ii) the numeric iterative methods depend on *good* starting points to converge and there are no general rules to find them for non-linear equations; (iii) attempting to solve the equations may reveal the non-uniqueness of the resulting geometry and the system has no basis for choosing 'the intended' one; (iv) consistency detection and maintenance in these systems are problematic since constraints are solved numerically and it is difficult for a designer to create and maintain a consistent set of constraints.

To avoid the problem of converting the constraints into a large system of equations, Rossignac [16] proposed an alternative approach based on the operational interpretation of constraints. Constraints are evaluated independently in a sequential manner and the sequence of evaluation is defined by the user. Two problems with this approach are (i) users are responsible for designing a sequence of operations to satisfy constraints and for resolving the conflicts in evaluation of the constraints, and, (ii) it is inadequate for solving problem formulated by several simultaneous constraints for which no simple sequential process is available.

Constraint propagation as an approach to satisfy a network of constraints has been studied by many researchers [5, 7, 19, 20]. The basic idea of constraint propagation is to compute a value whenever possible and propagate it to related constraints and repeat the process until nothing can be computed or a contradiction occurs. This works well for computing a set of assignments for a network of cells that must satisfy some algebraic constraints. Geometric constraints are usually translated into algebraic equations after which the problem becomes that of finding a set of assigments to a set of parameters. This, however, sacrifices the semantics of geometry and is inefficient when a large set of geometric constraints is concerned. It will be more desireable to propagate geometric constraints at higher semantic level.

As it was observed in [1], the analytic view of geometric constraints is not sufficient for high level applications like mechanical design. We also need to use the axiomatic view of geometry explicitly to represet and reason about geometric constraints at an abstract level. We are using an approach, called Operational Transformation Planning (OTP), which is based on the high level understanding of the semantics of constraints and the geometric implications of operations. The main ideas are (i) constraint satisfaction and maintenance can be carried out in terms of operations and (ii) a network of constraints can be satisfied by planning a sequence of operations through geometric reasoning.

Oar is an object oriented programming system where general declarative information about objects and their interrelationships dynamically determines their imperative behavior [3]. In Oar, logic programs determine what objects receive what messages and what procedures they use to respond to them. A procedure itself is a sequence of message passing imperatives that request the services of other objects. Oar programs can exercise or relinquish explicit control during the course of their computation in proportions suitable for the application at hand. Oar is especially suitable for highly dynamic applications such as design, where objects and their behavior must evolve throughout the course of a computation, as more information about them and their interrelationship is assimilated by a program. The dynamic, context-dependent inheritance of methods by objects and the symbolic reasoning capability supported by Oar make it a suitable environment for OTP.

The rest of the paper is organized as follows. §2 is a brief review of some related work. §3 is an overview of a prototype design system based on OTP from designer's point of view. The system architecture is introduced in §4. The principles of the OTP approach are presented in §5. In §6, an implementation scheme in Oar is presented. A case study is discussed in §7. §8 summarizes some of the important characteristics of the system. Finally, §9 concludes the paper.

2. Related Work

In 1962, Ivan Sutherland built the first constraint-based drawing system SKETCHPAD [21] that allowed the definition of arbitrary objects and constraints. It pioneered the use of interactive computer graphics and constraint systems. Alan Borning's THINGLAB system [5] carried on where SKETCHPAD left off. THINGLAB is a generalized simulation laboratory based on constraints. Users sketch a design and tell THINGLAB what the parts are and how they behave, and THINGLAB performs a simulation.

Variational geometry [8, 13-15] is a typical algebraic approach that represents and satisfies geometric constraints algebraically. The central idea of their effort is that dimensions are a natural descriptor of geometry and provide the most appropriate means for altering a geometric model. In variational geometry, a geometric model is defined with respect to a set of characteristic points in space. Dimensions are treated as constraints limiting the permissible locations of these characteristic points. Each dimensional constraint is described by a nonlinear equation involving the coordinates of associated characteristic points. A given dimensioning scheme is represented by a set of such equations. The geometry corresponding to an altered dimension is found through the simultaneous solution of the set of constraint equations. The Newton-Raphson method is used to simultaneously solve the constraining equations which yields the geometry corresponding to the set of dimensions. Some techniques are used to identify invalid dimensioning schemes and to improve the efficiency of solving equations.

G. Sussman and G. Steele defined and implemented a constraint programming language and explored the declarative nature of constraints using symbolic computation [19, 20]. James Gosling [7] investigated the algebraic constraints by algebraic transformations that transform a constraint graph into some manageable form. He implemented his algorithms in MAGRITTE, an interactive graphical layout system.

Bruderlin et al [6] developed a system for automatically building 3-D geometric objects that are defined by their topology and by geometric constraints. A 3-D solid modeller providing set operations on polyhedra is used as a graphic editor for sketching spatial objects and thus defining their topology. Sketched objects are dimensioned (constrained) by selecting points and entering values. The geometric constraints are first evaluated symbolically by a constraint satisfaction mechanism written in the logic programming language Prolog. The result of the symbolic evaluation is a set of construction steps. In the second step, the symbolic solution is then numerically evaluated by procedures written in the procedural programming language Modula-2.

Kimura and Suzuki [9, 10] presented a uniform framework for product modelling where geometric constraints capture a wide range of geometric information, such as dimensions, tolerances and assembly. They proposed a method that combines first order predicate logic and an object oriented approach to deal with geometric constraints. In their framework, they use solid models for storing the product's shape description and logical formulae for representing geometric constraints on the elements of the solid models. Dimensions and tolerances are represented explicitly as logical formulae. Geometric reasoning (logic inferencing) is a process which manipulates these formulae to generate new formulae and to find out a solution satisfying a set of constraints.

Arbab and Wing [1] explore how axiomatic knowledge of geometry can be incorporated into geometric modellers to enhance the semantics of representation and manipulation of geometric information. They consider geometric problem solving as a process that involves the axiomatic view of geometry as well as its analytic view. A programming environment, Oar (Object and Reasoning) [4] was developed, which combines logic programming and object oriented programming to accommodate geometric reasoning.

Rossignac [16, 17] proposed an approach to facilitate specification and editing parameterized models of solids through a user-friendly interactive graphical front-end in systems based on a dual representation. The main idea is to specify positions and dimensions of primitives in terms of geometric constraints but not through parameters of rigid motions. In his approach, a user specifies a binary CSG graph and a sequence of motions that operate on collections of primitives. The rigid motions are stored in terms of unevaluated constraints on graphically selected boundary features, the constraints are evaluated sequentially in a user-specified order.

Recently, Serrano and Gossard presented a constraint-based environment MCAE, for Mechanical Computer Aided Design [18]. Constraint networks are represented as directed graphs, where nodes represent parameters and arcs represent constraint relationships. They developed a bipartite matching representation to generate the parameter dependencies automatically. They use algorithms based on the constraint network topology to evaluate constraint networks, detect over- and under-constraint systems, and identify and correct redundant and conflicting constraints.

3. System Environment

We are building a prototype constraint-based design system for the domain of two-dimensional geometry using OTP. As the front-end of this system, a graphical interface is built that provides users with means to interact with the system. The interface provides a sketch pad and a list of command icons. A user can experiment his design by picking the command icons and position points on the sketch to create objects, modify objects or impose constraints.

Command icons are organized hierarchically. Two main classes of icons are *to create objects* and *to impose constraints* (abbreviated as commands CREATE and CONSTRAIN). CREATE consists of a list of commands to create an instance of various classes of objects recognized by the system. When CREATE is picked, its list of command icons is displayed at the bottom of the sketch pad. A user can create an instance of any object by clicking the icon of its corresponding object type and picking the position points for the instance. A user is required to enter a character string as the name of the object created which is used to refer to this object later. For example, a user can create a line by clicking LINE and picking its two end-points and entering l_1 as its name. When this is done, a line is displayed in the sketch (as in Fig. 1). An instance of a circle can be constructed similarly by picking its center and a point on its circumference. The construction process is guided by the system by prompting for user input. A user can end the creation phase and returns to the higher command level by clicking the QUIT icon. Additional commands are provided by CREATE, such as to delete an object and to *read in* objects previously created in a file.

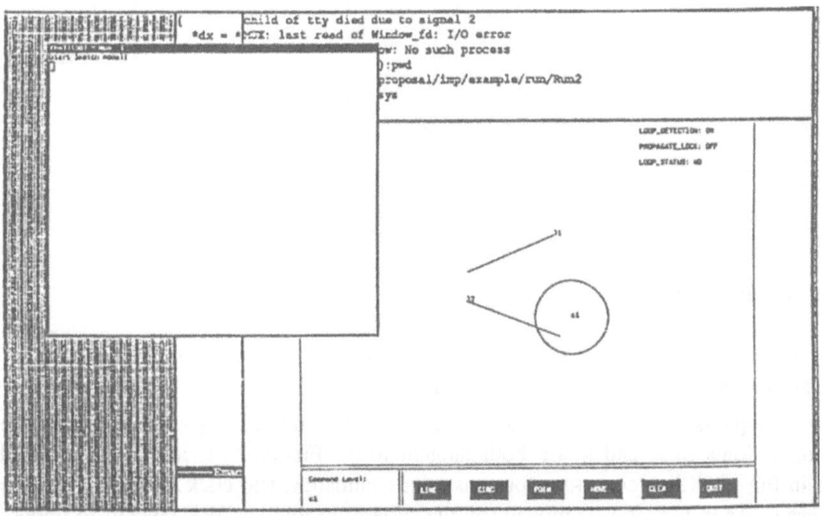

Fig. 1. Two unconstrained lines and a circle

The second class of icons, under CONSTRAIN, contains a list of commands to impose a constraint of various types recognized by the system. When CONSTRAIN is clicked, the list of command icons is displayed at the bottom of the sketch pad. A user can impose a constraint of any type by clicking its corresponding icon and picking the constrained objects in the sketch. A user is required to enter a name for the constraint which will be used for reference either in satisfying a constraint or maintaining a constraint. For example, to impose the constraint l_1 join l_2, a user can click icon JOIN, pick l_1 and l_2 on the sketch and enter con_1 as the constraint name. When done, l_1 and l_2 join at their end-points (as in Fig. 2). A user can end the constraint phase and return to the higher command level by clicking the QUIT icon.

Another important command MOVE is used to *move* existing objects around the sketch pad. MOVE can be interpreted in a generic sense as a combined translation and rotation involving the positions of objects only. To move an object, a user simply clicks the MOVE icon and picks an object in the sketch, then picks a point where this object is moved to. When done, the object is redisplayed in the new position. The execution of the MOVE may result in violation of satisfied

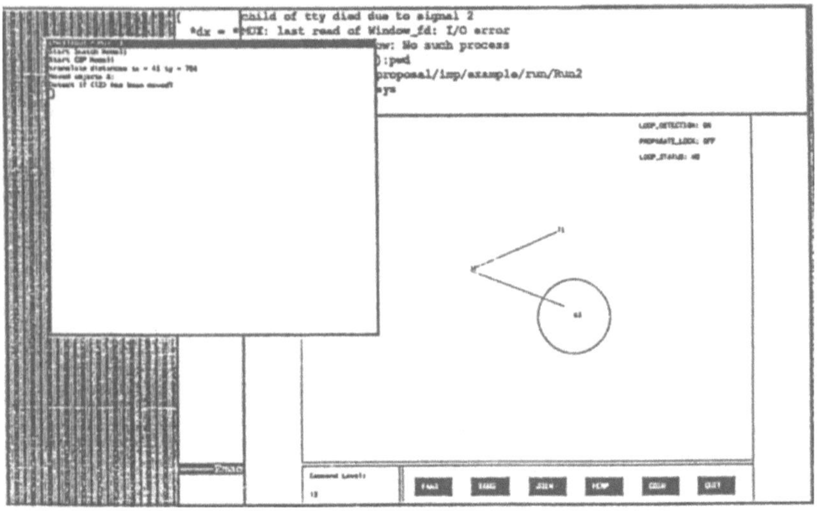

Fig. 2. Joined lines and a circle

constraints, therefore some other changes may be made by the system to maintain the constraints. The details are discussed in later sections.

A simple example is demonstrated below which involves two lines l_1, l_2 and a circle c_1. The constraints are: l_1 joins l_2, l_1 and l_2 are both tangent to c_1. Firstly, l_1, l_2 and c_1 are created separately as in Fig. 1. Click CONSTRAIN for constraint commands, and click JOIN to impose the constraint l_1 join l_2. As in Fig. 2, l_2 is moved (by the system) to join l_1. Click TANGENT to impose constraint l_1 tangent c_1, as in Fig. 3, c_1 is moved (by the system) to be tangent to l_1. Imposing the third constraint l_2 tangent c_1, instead of moving c_1 to satisfy the constraint, results in the system rotating l_2 around the joint to become tangent to c_1 as in Fig. 4. This is done through reasoning about the context of the satisfied constraints.

4. System Architecture

The system consists of three major components: the user interface, the sketch model, and the constraint model (Fig. 5). Internally, the three components are objects and communicate with each other through message passing. The sketch model is an internal representation of the geometric information displayed on the screen. The constraint model is an abstract representation of the constraints imposed by a user. A user can directly manipulate both models. Unlike most other constraint systems, the constraint model is not necessarily *complete* enough to imply a unique solution. Specifically, the loose binding between the two models means that the sketch model is always only an instance of a solution for the constraint model. Thus, adding a constraint may result in a modification to the sketch model, while deleting a constraint causes no change to the sketch model. Similarly, modifications to the sketch model may or may not entail

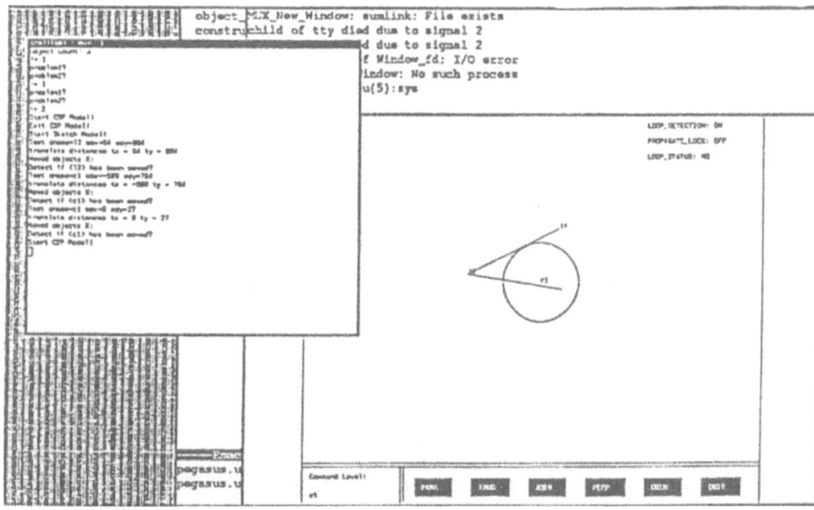

Fig. 3. Circle tangent to one of the lines

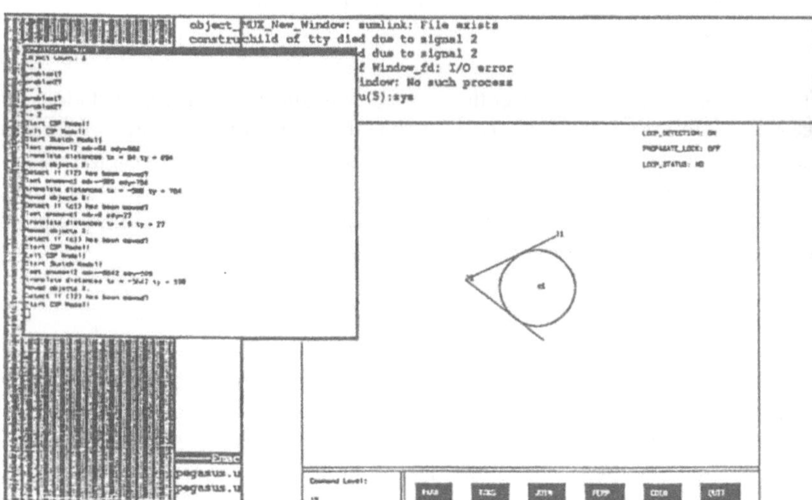

Fig. 4. Circle tangent to both lines

further changes to the sketch in order to keep the sketch model an instance of the constraint model.

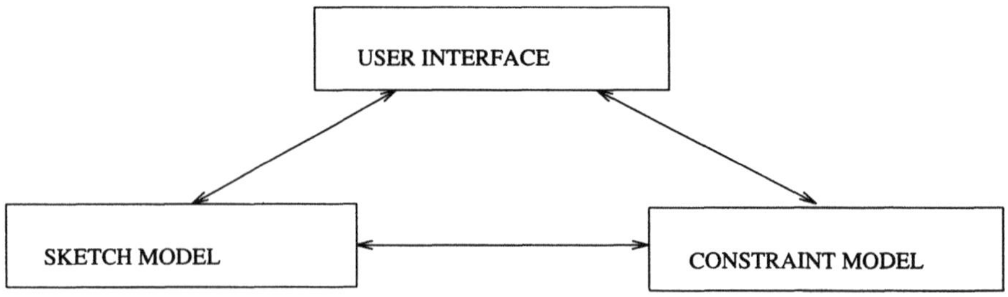

Fig. 5. Top level architecture of the system

User interface

The user interface is a graphical window with command panes by means of which a user can interact with the system. Basically, the user interface acts as a command interpreter which interprets users' input and passes proper messages to either the sketch model or the constraint model. Users can change the state of the system only through the sketch window and command panes. Simultaneously, in the other direction, the user interface also interprets system's response (messages from the sketch model or the constraint model) on the sketch.

Sketch model

The sketch model is essentially a system object that manages the dimensioned representations of objects in the sketch window. It has methods corresponding to messages for creating objects and modifying objects. It deals with individual object instances in terms of their dimensional values in the sketch. It can also communicate with the constraint model via message passing when changes are made to the sketch or when a constraint is imposed. The sketch model may request the user interface to redisplay objects whose dimensional values are changed.

Constraint model

The constraint model is conceptually a system object that maintains the symbolic information about the objects in the sketch. It represents the properties of the objects and their relationships (constraints) symbolically, and has an inference mechanism to reason about this information. The constraint model has methods to satisfy constraints and to maintain (previously satisfied) constraints. The constraint model responds to messages from the sketch model by invoking its methods to satisfy constraints or to maintain constraints. Geometric reasoning is used to choose the proper methods.

5. Operational Transformation Planning Approach

We use operational transformation planning (OTP) as an approach to satisfy a network of geometric constraints incrementally. The main ideas are as follows. A constraint is satisfied by performing a sequence of operations specifically designed for this constraint. A satisfied constraint can either tolerate or propagate an operation performed on its participant objects. Satisfying a constraint in the context of a network of satisfied constraints is to perform the operations associated with the constraint and to propagate the changes through the network in term of operations. The satisfaction process is planned through symbolic reasoning. A set of constraints is satisfied incrementally.

5.1. The computational model

The operational transformation planning model is a 5-tuple $<\mathcal{O},\mathcal{C},\mathcal{P},\mathcal{R},\mathcal{I}>$. $\mathcal{O}$ is a finite set of *object types* that contains all object types of interest. An object type actually defines a class of *object instances*. An *object instance* of an object type (abbreviated as *object*) is a concrete entity with dimensional values.

$\mathcal{C}$ is a finite set of *constraint types* that specify various relationships among participating object types. A *constraint type* $c_i(o_1, o_2, ..., o_r)$ consists of two parts: constraint name c_i and a list of participating object types $<o_1, o_2, ..., o_r>$. A *constraint type* defines a class of *constraint instances*. A *constraint instance* (or *constraint*) is a constraint type where its participating object types are replaced correspondingly with object instances. These object instances are called *participants* in the constraint.

$\mathcal{P}$ is a finite set of *operation types*. An *operation type* $p_j(o_1, o_2, ..., o_s, \phi_1, \phi_2, ..., \phi_t)$ consists of three parts: operation name p_j, a list of object types $<o_1, o_2, ..., o_s>$ and a list of parameters $<\phi_1, \phi_2, ..., \phi_t>$. An *operation type* defines a class of *operation instances* (abbreviated as *operation*). An *operation* is an operation type where its participating object types are replaced correspondingly with object instances. Operations are applied to object instances in order to (1) satisfy a new constraint imposed on them, and (2) maintain the already satisfied constraints they participate in. An operation can be as simple as a primitive translation or a rotation, or it may be some complex procedure for finding the clipped version of a polygon.

$\mathcal{R}$ is the constraint propagation function used to maintain the satisfied constraints in a network after a change. A satisfied constraint $c_i(a_1, a_2, ..., a_k, ..., a_r)$ receiving an operation $p_j(b_1, b_2, ..., b_s, \phi_1, \phi_2, ..., \phi_t)$ from its kth participant, fires a set of operations defined by $\mathcal{R}(c_i, k, p_j)$. This function returns alternative sets of pairs $\{<p_l,n_l>\}$. A set of pairs $\{<p_l, n_l>\}$ indicates that in order for the constraint c_i to *tolerate* operation p_j on its kth participant, operation p_l must be performed on its n_lth participant for each pair in the set.

$\mathcal{I}$ is the impose function used to impose a new constraint c_i on a set of objects that may already participate in a network of satisfied constraints. The function $\mathcal{I}(c_i)$ returns alternative sets of pairs $\{<p_l, n_l>\}$. A set of pairs $\{<p_l, n_l>\}$ indicates that to satisfy c_i, operation p_l must be applied to its n_lth participant for each pair in the set.

5.2. Satisfaction planning

Constraint satisfaction in the OTP approach is actually carried out in two stages: (1) planning a sequence of operations and (2) executing the operations. A set of constraints is satisfied incrementally by repeating the satisfaction process.

At the planning stage, when a new constraint c_i is added, it sends the operations prescribed by $\mathcal{I}(c_i)$ to its participants. Objects that are participants in other (satisfied) constraints in the network, notify them of the proposed operations. A constraint receiving an operation from one of its participants may send some other operations to its other participants as prescribed by its $\mathcal{R}$ function. A participant is *isolated* if it participates in only one constraint. An isolated participant can *absorb* any operation that maintains the constraint. A constraint *absorbs* an operation from a participant if the function $\mathcal{R}$ returns an empty set of operations. If this propagation of operations in the network terminates in a finite number of steps, the resulting sequence of operations is a plan for satisfying the new constraint in the context of the previously satisfied constraints. The second stage is the execution of the plan. Specific procedures are associated with the entries in the $\mathcal{R}$ and I tables to carry out the operations.

Geometric knowledge and the context information of constraints can be incorporated in the planning stage by means of geometric reasoning. There are two cases when alternative schemes of imposing or propagating operations exist and we should take the best choice. We can also take advantage of the degrees of freedom in choosing parameters of operations, using the context of satisfied constraints and geometric knowledge. For example, to satisfy a constraint l_1 parallel l_2, we can rotate l_1 by a certain angle θ around any pivot, but if the context is that l_1 tangent to another circle c_1, we would like to rotate l_1 around the center of c_1, so that we preserve tangency. As another example, to make l_1 tangent to circle c_1, in the context l_1 intersecting l_2, we would prefer to rotate l_1 around the intersection point rather than to translate l_1.

5.3. Propagation loops

The propagation of operations during a satisfaction planning process may result in a loop which indicates the constraints involved cannot be dealt with in isolation. Global relaxation of the whole constraint network as used in THINGLAB [5] is a general way to solve the loop problem but it is inefficient. We present a solution using local relaxation which solves only those constraints involved in a loop and propagates the solution to the rest of the constraint network. A simple example below illustrates the loop problem.

The pentagon problem

Given five segments s_1, s_2, s_3, s_4, s_5 with fixed lengths, construct a pentagon on a plane. A simple OTP model for this problem is given below.

$O = \{$ segment,point $\}$.
$C = \{$ joint(segment,segment,point,point) $\}$.
$\mathcal{P} = \{$ link(segment,segment,point,point) $\}$.
$\mathcal{R}(\text{joint}(S_1,S_2,P_{e1},P_{s2}),\text{link}(S_3,S_1,P_{e3},P_{s1})) = \text{link}(S_1,S_2,P_{e1},P_{s2})$.
$I(\text{joint}(S_1,S_2,P_{e1},P_{s2})) = \text{link}(S_1,S_2,P_{e1},P_{s2})$.

The constraint joint(S_1,S_2,P_{e1},P_{s2}) means that the starting point P_{s2} of segment S_2 coincides with the end point P_{e1} of segment S_1.
The operation link(S_1,S_2,P_{e1},P_{s2}) moves segment S_2 such that its starting point P_{s2} coincides with the end point P_{e1} of segment S_1.
The pentagon can be constructed by a set of five constraints: {joint(s_1,s_2,p_{e1},p_{s2}), joint(s_2,s_3,p_{e2},p_{s3}), joint(s_3,s_4,p_{e3},p_{s4}), joint(s_4,s_5,p_{e4},p_{s5}), joint(s_5,s_1,p_{e5},p_{s1})}.

The first four constraints can be satisfied using the OTP approach without any problems. The fifth constraint joint(s_5,s_1,p_{e5},p_{s1}) which tries to close the pentagon results in a loop. The impose function I(joint(s_5,s_1,p_{e5},p_{s1})) returns {link(s_5,s_1,p_{e5},p_{s1})}. As in Fig. 6, the movement of s_1 involves the movement of s_2, and the movement of s_2 involves the movement of s_3, etc., and finally, the movement of s_5 involves the movement of s_1: the propagation of the link operation forms a loop.

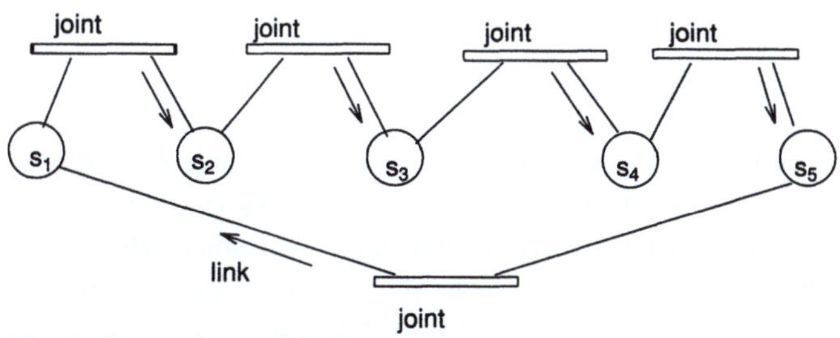

Fig. 6. Constraint model of a pentagon

Local relaxation

Local relaxation handles a set of constraints which cannot be satisfied in isolation. The main steps are i) detect a propagation loop in the current propagation phase; ii) translate the constraints involved in the loop into a system of algebraic equations; iii) pick up the additional dimensional equations from the sketch model that are necessary to make the set of equations solvable; iv) solve the system of equations numerically; v) assemble a sequence of operations according to the numerical solution, so that when the operations are carried out, the constraints in the loop are satisfied simultaneously; and finally, vi) propagate the assembled operations to the rest of network. We briefly explain the above steps through an example.

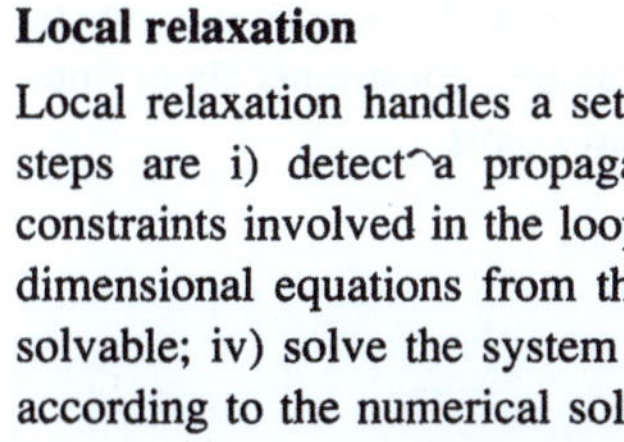

Fig. 7. Making a constrained pentagon

Consider the line segments in Fig. 7. The underlying constraints are l_1 parallel l_2, l_1 parallel s_1 and four joint constraints that connect the five segments s_1, s_2, s_3, s_4 and s_5. To close the five segments, we impose a new constraint s_1 joint s_5 which results in a loop.

1. Detection of Propagation Loops

   The propagation of operations originates from satisfying a constraint. A constraint sends out operations to its participants either to satisfy or maintain itself. An object involved in the constraint receives the operation and propagates to other constraints it participates in. The underlying assumption is that the objects receiving an operation have not received any

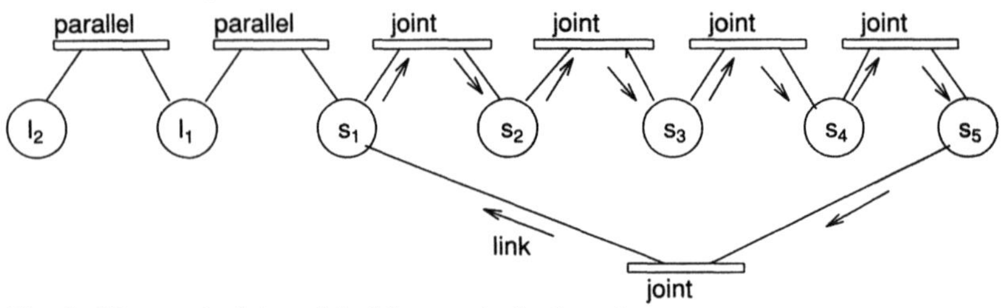

Fig. 8. The constraint model of the constrained pentagon

other operations during the current satisfaction planning process. If an object receives a second operation, it contradicts our assumption of operation. We call it a propagation loop.

In this example, a loop is detected when s_1 receives the second link operation. The rest of the task is to find all the constraints involved in the loop. This is done by backtracking the propagation route. In the example, the five joint constraints are collected.

2. Translating Constraints into Equations

The set of constraints (and objects) obtained from the above step is translated into a system of algebraic equations. The equations are in terms of coordinates of characteristic points or characteristics parameters of the participating objects. For example, constraints about lines are translated into equations in terms of coordinates of their end-points.

l_1: $(x_{11},y_{11}), (x_{12},y_{12})$
l_2: $(x_{21},y_{21}), (x_{22},y_{22})$

joint$(l_1,l_2) \rightarrow x_{21} = x_{12},\ y_{21} = y_{12}$
parallel$(l_1,l_2) \rightarrow (y_{12}-y_{11})(x_{22}-x_{21}) = (y_{22}-y_{21})(x_{12}-x_{11})$

An initial set of equations are obtained from the constraint model. The resulting system of equations is usually under-constrained, because the constraint model generally does not specify the positions and orientations of all objects. We need to pick up additional constraints from the sketch model to make the system of equations solvable. This gives us some freedom in choosing additional dimensional constraints from the sketch model. For instance, we simply fix some objects to obtain a well-constrained set of equations. We can apply heuristic rules to select an appropriate set of additional constraints to 1) localize the effect of the new constraint, and 2) simplify equation solving. In this example, we would like to fix s_1 that maintains the parallelism between l_1 and s_1.

3. Solving the System of Equations

The system of equations is solved using numerical relaxation methods. We use a method to represent and manipulate equations similar to the one used in variational geometry. The solution gives the coordinates of the characteristic points or parameters of the objects involved in the loop.

4. Assemble Operations From the Solutions

   The solutions of equations give the proper dimensional values of the objects that satisfy the constraints in the loop. Objects whose original positions are different from the solutions must be moved to their new positions by some operations. The effect of such operations may have to be propagated to the objects outside of the loop. So we assemble the operations according to the solutions and propagate the operations to the rest of the network. The assembling of operations depends on the context of the rest of network. When alternatives exist, we choose the operation which affects the rest of network less than others.

5. Propagate Operations to the Network

   Every one of the operations may propagate to the rest of the network. In this example, l_1 parallel s_1 is maintained because s_1 is fixed in loop relaxation and no further propagation is required. Otherwise, the operations on s_1 must be propagated to l_1 through parallel to maintain the constraint l_1 parallel s_1.

6. Constraint Management

6.1. An overview of Oar

Oar is a system for object-oriented programming that combines the power of logic programming for expressing declarative facts about objects and their interrelationships, with the concept of message passing as the mechanism for triggering the imperative knowledge associated with objects [3]. In Oar, general declarative information about objects and their interrelationships determine an object's behavior. The details of manifestation of such behavior are expressed imperatively through message passing. The inheritance network in Oar is dynamic and fluid. The class membership of an object, and thus its behavior, can dynamically change as more facts about its properties and its relationships with other objects are introduced. Oar is suitable for highly dynamic applications where objects and their behavior must evolve through the course of a computation, as more information about them and their interrelationship is assimilated by a program.

Computation in Oar is carried out by two major components: an inference mechanism based on first order logic, and a imperative programming environment based on message passing. The two components are closely interrelated through the notion of objects. Message passing is the means by which objects use the computational resources of other objects. Message passing in Oar is associative and uses the inference mechanism to identify the receivers of messages and their methods. The inference mechanism itself is an object whose resources can be used by other objects through message passing. Some of the interesting features of Oar are presented below.

1. Combining declarative and imperative knowledge

   In most object oriented systems, declarative knowledge can be specified for only one purpose. This is to reflect a limited view of how objects are interrelated through a class/instance hierarchy. As in logic programming languages, in Oar, declarative knowledge of a general nature can be expressed. Such information can affect selection and activation of methods of objects.

2. Predicates as object properties

   In a language like Smalltalk, an object is a concrete entity whose structure is determined (and fixed) at the time it is instantiated, by its position in a predetermined class hierarchy. Thus, structural properties of objects, i.e., their attributes, are induced by this static class hierarchy, although their values may change dynamically. As in logic programming languages, there is no such distinction between structure and value in Oar. In Oar, an object is an abstract entity that satisfies a given set of properties. Properties of an object in Oar are the set of facts (well-formed formulae in a formal system based on first order logic) known about that object. The properties of an object include any well-formed formula that can be derived by the inference rules of the logical system, as well as the ones stated explicitly, i.e., as axioms.

3. Dynamic object hierarchy

   A distinguished relationship among objects in Oar is the **isa** relation. This relation defines an object as a specialization of another. An **isa** relation between two objects can be defined as the logical consequence of some other facts. Thus, in Oar the equivalent of a class hierarchy is induced by the properties of objects. Because properties of objects can dynamically change as more facts are discovered during program execution, the object/class hierarchy in Oar is dynamic.

4. Prototype objects

   In Smalltalk, there is a distinction between an object and a class. The underlying philosophy of this distinction is the set representation of a class [11]. A class is a set of instances where each instance is an object. Classes can have subclasses. Subclass definition and creating an instance are two distinct operations in this approach. In Oar, a class is represented by one of its typical members, i.e. by an object. This approach is based on the representation of classes as *prototypes*. In this approach, any object represents the class of which it is a typical member. Defining a subclass and creating an instance are both accomplished by the same mechanism: defining extension objects through the **isa** predicate.

5. Inheritance of properties

   Inheritance in Oar is associative. Propagation of properties through the object/class hierarchy (Oar's equivalent of inheritance in Smalltalk) is accomplished through a special inference rule: the **isa** inference rule (a different concept than the **isa** relation mentioned above). The combination of this inference rule and Oar's method definitions make message passing in Oar effectively similar to the delegation mechanism of actor-based systems [12].

6. Associative binding of methods with objects

   Associative inheritance of method properties leads to a flexible and powerful generalization of the binding of imperative knowledge with objects. In most object oriented programming languages, the binding of methods with objects is static and by programmers' decree. In Oar, this binding is associative and dynamic. For instance, It is possible to define a method as applicable to an abstract (class of) object(s) whose properties satisfy the preconditions of the procedure. Thus, any object whose properties satisfy the preconditions of such a method, will dynamically inherit the method. Furthermore, this ensures that the method is never invoked by an object that does not satisfy its preconditions.

7. Associative message passing

   Message passing in Oar is a generalization of message passing in conventional object oriented languages. To send a message, an Oar object specifies a message and a qualifier. The informal semantics of message passing in Oar is that the sender activates the specified method of all objects in the closure of the given qualifier. This generalizes the conventional message passing in two ways: first, there can be more than one receiver for a message, and second, the sender does not necessarily directly know the receivers beforehand.

Associative inheritance, dynamic object hierarchy, and associative message passing create alternative courses of action. When the message passing mechanism finds more than one receiver for a message, or more than one method for a receiver, they are taken as representing alternative courses of action. One of these alternatives can be selected nondeterministically, or they all can be pursued in parallel in a manner that is transparent to the program, through *branching*. The user program can also take control and sequentially send a message to all receivers as in a broadcast. For example, a user program may want to send a message to all objects whose color is red and let them print their names sequentially.

6.2. Mapping an OTP model into Oar

Object types

Objects of an OTP model are Oar objects represented as object symbols. In Oar, there is no distinction between an object type and an instance of the type. An object type is a prototype object of that type.

The object types of an OTP model are represented as a set of object symbols reserved for the types of objects recognized by the system. An instance of a type can be declared using the **isa** predicate. For example, line is an object symbol representing object type line, a specific line l_1 can be declared as:

l_1 **isa** line.

Through the **isa** relation, l_1 can inherit all properties of line.

Constraint types

Constraint types are also Oar objects, therefore represented as object symbols as well. Same as object types, there is no distinction between a constraint type and an instance of the type. A constraint type is a prototype constraint of that type.

The constraint types of an OTP model are represented as a set of object symbols reserved for the constraint types recognized by the system. An instance of a type can be declared using the **isa** predicate. For example, parallel is an object symbol representing constraint type parallel, a specific constraint p_1 can be declared as:

p_1 **isa** parallel.

Operations

An operation of an OTP model is implemented as a method procedure and is denoted by an object symbol which is the (message) selector for the method procedure. Operations are always associated with object types as their methods. The association is specified by the **method** predicate. For example, the association between the object type line and its method rotate which is implemented as the method procedure line_rotate_c can be specified as:

```
method(line,rotate,2,line_rot_c).
```

This definition states that when the object line, or any of its specializations that inherits this definition, receives a message whose selector is rotate and contains 2 parameters (a pivot and an angle), the method procedure line_rot_c is to be invoked to respond to the message.

An object type can have several method procedures for the same message selector. Selection of a specific method procedure to respond to a message can depend on the context. For example,

```
method(line,rotate,2,line_rot1_c):-
        tangent(line,circle).
```

states that when a line is tangent to a circle, the method procedure line_rot1_c is to be invoked to respond to a message whose selector is rotate and has 2 parameters. The procedure line_rot1_c is implemented to rotate a line around a circle and line_rot_c rotates the line around its start point. The selection of a method procedure can be based on by preconditions required by the method clause. For instance, line_rot1_c requires the line to be tangent to a circle. Association of the method procedures is determined by the Oar's inference mechanism dynamically at run time.

The general format for an operation method procedure is as follows:

```
procedure_name(client,selector,par1,par2,...,parn)
{
        check_operation_loop;
        perform_operation;
        propagate_operation;
}
```

Impose Function

The impose function is distributed to each constraint type as methods to satisfy the constraints. In our prototype implementation, the common selector for these methods is satisfy. Each constraint type has a method procedure that implements the impose function on this constraint type. The alternatives returned by impose function are implemented as different method procedures which are selected depending on the context. For example, the definitions

```
method(tangent,satisfy,2,tangent_sat1_c):-
        constrained_by(Line,tangent),intersection(Line,Line1).

method(tangent,satisfy,2,tangent_sat2_c).
```

states that tangent_sat1_c is invoked to satisfy an instance of tangent, if the line constrained by (the specific instance of) tangent is already constrained by an intersection constraint. This procedure rotates the line around the intersection such that the line is tangent to the circle. Otherwise tangent_sat2_c is invoked which simply translates the line to be tangent to the circle. The association of the method procedures and objects is dynamic at run time.

The general format of a method procedure for the method satisfy is:

```
procedure_name(client,selector,par1,par2,...,parn)
{
        prove_by_theory; /* geometric reasoning */
        numerical_checking;
        parameter_computation;
        operation_sending;
}
```

Propagation Function

The propagation function is distributed to each constraint type as methods to maintain the constraints. In our prototype implementation, the common selector is maintain. that implements the propagation function on this constraint type. The alternatives returned by the propagation function are implemented as different method procedures specifying the context-dependent conditions for their selection. For example, the definitions

```
method(tangent,maintain,2,tangent_mai1_c):-
        constrained_by(Line,tangent),intersection(Line,Line1).

method(tangent,maintain,2,tangent_mai2_c).
```

states that tangent_mai1_c is invoked to maintain an instance of tangent, if the line constrained by tangent is also constrained by another intersection constraint. For example, to maintain the constraint under an operation that translates the circle away, the procedure rotates the line around the intersection to keep the line tangent to the circle. Otherwise tangent_mait2_c is selected which simply translates the line to be tangent to the circle.

The general format of a method procedure for method maintain is as follows:

```
procedure_name(client,selector,par1,par2,...,parn)
{
        parameter_computation;
        operation_sending;
}
```

6.3. Geometric reasoning

Geometric reasoning can be used in OTP to make the process of planning for constraint satisfaction more intelligent. As we observed in previous sections, the satisfaction planning process is rather local and simple-minded. More intelligent plans can be generated by considering the context of a constraint and its participating objects.

For example, suppose line l_1 is tangent to circle c_1. Imposing the constraint l_1 parallel l_2, can result in performing a rotation on l_1. Normally, rotation of a line would use its start point as the pivot. We can associate a different rotation method procedure with l_1 to use the center of the circle as the pivot for the rotation, so that the tangency is preserved.

```
method(line,rotate,2,object_smart_rotate):–
    constrained_by(line,tangent).

method(line,rotate,2,object_normal_rotate).
```

The context sensitivity intends to short-cut the operation propagation so that the possibilities of operation loops are reduced. In general, we try to localize the modification of the satisfied network and therefore reduce the amount of computation.

The explicit representation of geometric knowledge makes the reasoning about geometric constraints possible. The following example shows that implied constraints are satisfied without operational transformation.

If l_1 parallel l_2 and l_2 parallel l_3 are two satisfied constraints, imposing l_3 parallel l_1 is redundant, the system can prove this is an implied constraint and does nothing to satisfy it. The geometric knowledge used can be represented as:

```
method(parallel,satisfy,2,do_nothing):-
    constrained_by(L1,parallel),
    constrained_by(L2,parallel),
    parallel(L1,L3),
    parallel(L2,L3).
```

Implied constraints always cause operation loops if they are not detected. Similarly, some contradictory constraints can be detected. Imposing l_1 perpendicular l_2 is a contradiction if we can show l_1 and l_2 are not perpendicular, for instance, show l_1 and l_2 are parallel.

```
method(perpendicular,satisfy,2,report_contradiction):-
    constrained_by(L1,perpendicular),
    constrained_by(L2,perpendicular),
    not_perpendicular(L1,L2).
```

6.4. System dynamics

1. *Constraint Satisfaction*:

   Constraint satisfaction is started by sending a message to a constraint object to invoke its satisfy method. The specific satisfy method procedure is dynamically selected by the system considering the context of the constraint. It sends proper messages to the constrained objects to invoke their operation methods. The operation procedures are also selected in a context dependent manner and change the objects properly and propagate the operations to other related constraints by invoking their maintain methods. The satisfaction process is done when the propagation finishes successfully or when a loop is detected.

2. *Loop Detection and Handling*:

   A special system object called history is responsible for detecting propagation loops. Same as other objects, history has methods that handle various cases. Two major methods are record which records an operation to be propagated into the history fact base and loop_detect which checks the fact base to see if a loop occurs.

When an operation method procedure of an object is invoked, it first sends a message to history to invoke its loop_detect method. If no loop is detected, the operation procedure performs the operation and sends a recod message to history. When a loop is detected, the loop handling process is invoked and the operation propagation stops.

Various specialized techniques can be used in the loop handler. In its most general form, the loop handler translates the constraints involved in a loop into a system of algebraic equations. Additional constraints can be picked up from the sketch model as needed to make the system of equations solvable. Using the solutions of the system of equations, the loop handler assembles a sequence of operations and sends them to the objects involved in the loop.

7. A Case Study

The following is a scenario that shows how a design problem is solved on our constraint system. The illustration is divided into steps that reflect the interactions between a designer and the system.

The Design Problem

Given two segments L, L1 with lengths l, l_1 and a circle C with radius r, design a mechanism which satisfies the following constraints:

(1) one end-point of L1 is on L;

(2) the other end-point of L1 is on C;

(3) the center of C is on the extension of L;

(4) the end-point of L1 on C can move around the circumference of C while sliding the other end-point on L and maintaining the length of L1.

The solution to this problem involves finding an appropriate position for L relative to C.

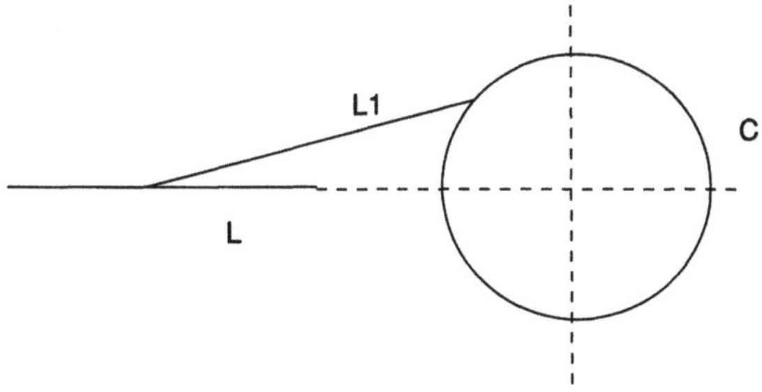

Denote the start point of a line L as $L_{startpoint}$, the end point of the line as $L_{endpoint}$ and extension of the line as $L_{extension}$. Denote the center of a circle C as C_{center}.

DESIGNER:

create an arbitrary line named L.
impose constraints: length(L) = l, horizontal(L).

SYSTEM:

scale L to length l.

the impose function for horizontal(L) returns:

rotate($L, L_{startpoint}, \phi$)

meaning to rotate L around its starting point by a proper angle ϕ such that L is horizontal.

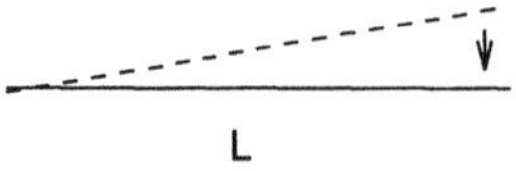

DESIGNER:

create another line named L1.

impose constraints: length(L1) = l_1, coincide($L1_{startpoint}$, L).

SYSTEM:

scale L1 to length l_1.

the impose function for coincide($L1_{startpoint}$, L) returns:

translate(L1, d_x, d_y)

meaning to translate L1 by a proper distance (d_x, d_y) such that its starting point is on L.

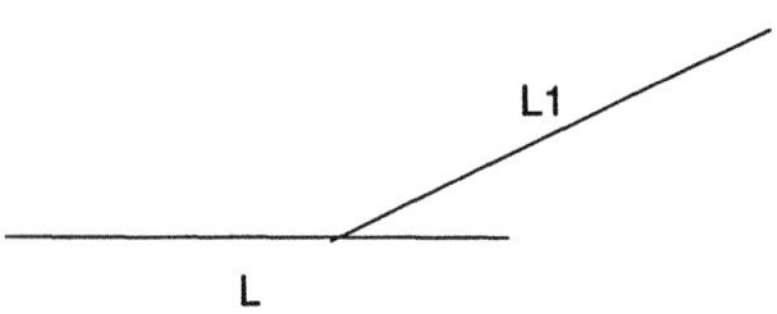

DESIGNER:

create a circle named C.

impose constraints: radius(C) = r, coincide(C_{center}, $L_{extension}$).

SYSTEM:

scale the radius of C to r.

the impose function for coincide(C_{center}, $L_{extension}$) returns:

translate(C, d_x, d_y)

meaning to translate C by a proper distance (d_x, d_y) such that the center of C is on the extension of L.

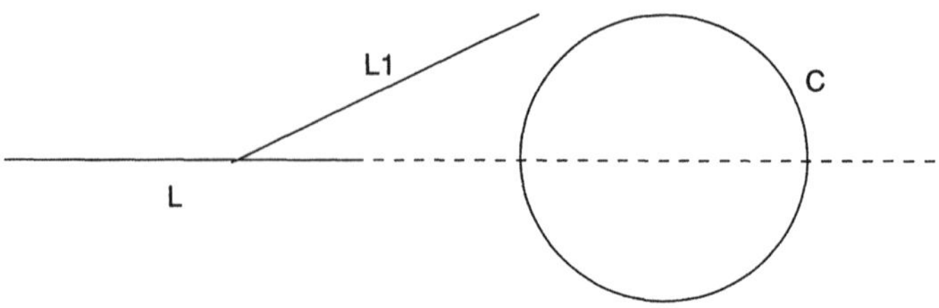

DESIGNER:

impose constraint: coincide($L1_{endpoint}$,C).

SYSTEM:

in the context of the constraint: coincide($L1_{startpoint}$,L),
the impose function for coincide($L1_{endpoint}$,C) returns:

if distance(C_{center},$L1_{startpoint}$)-radius(C) $\leq$ length(L1)
rotate(L1,$L1_{startpoint}$,ϕ)
else
translate(L1,L,d_{min})&rotate(L1,$L1_{startpoint}$,ϕ)

translate(L1,L,d_{min}) means translate L1 along L by the minimum required distance.

for the first case, rotate(L1,$L1_{startpoint}$,ϕ) is sent to the constraint coincide($L1_{startpoint}$,L) which decides it can tolerate this operation and no further propagation is required.

for the second case, translate(L1,L,d_{min})&rotate(L1,$L1_{startpoint}$,ϕ) are sent to the constraint coincide($L1_{startpoint}$,L).

depending on the amount of translation d_{min}, either both operations are tolerated by this constraint, or a translate(L,L,d_{min}) is propagated to L.

Assume it is the first case:

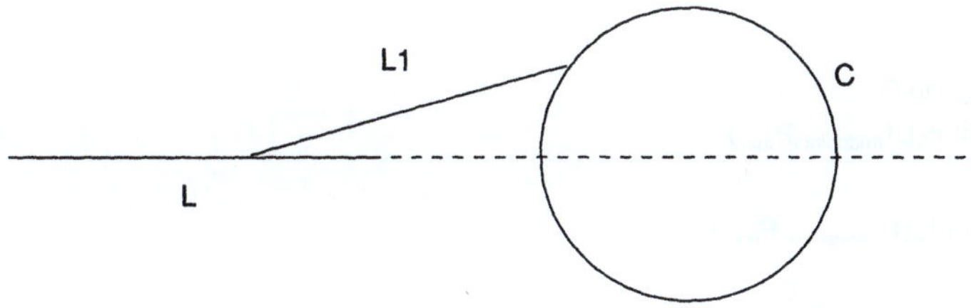

DESIGNER:

pick up a point P_{max} on C.
impose constraint: coincide($L1_{endpoint}$,P_{max})

SYSTEM:

in the context of the constraint: coincide($L1_{startpoint}$,L),
the impose function for coincide($L1_{endpoint}$,P_{max}) returns:
translate(L1,L,d)&rotate(L1,$L1_{startpoint}$,ϕ)

translate(L1,L,d)&rotate(L1,$L1_{startpoint}$,ϕ) are sent to the constraint coincide($L1_{startpoint}$,L).

depending on the amount of translation d, either both operations are tolerated by the constraint, or translate(L,L,d_{min}) is propagated to L, where d_{min} is the minimum distance required for L to move in order to keep coincide($L1_{startpoint}$,L) satisfied.

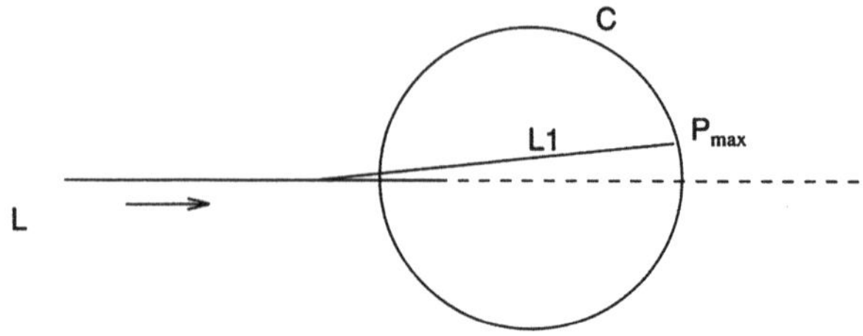

DESIGNER:

remove constraint: coincide($L1_{endpoint}$,P_{max}).

SYSTEM:

remove coincide($L1_{endpoint}$,P_{max}) from constraint set, nothing is changed in the sketch.

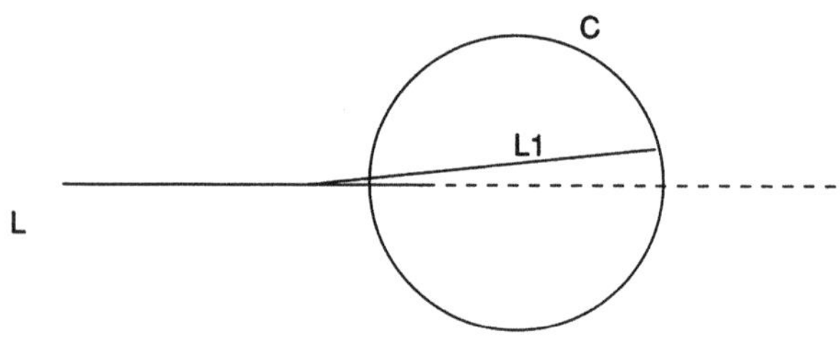

DESIGNER:

pick up another point P_{min} on C
impose constraint: coincide($L1_{endpoint}$,P_{min})

SYSTEM:

similar to imposing coincide($L1_{endpoint}$,P_{max}).

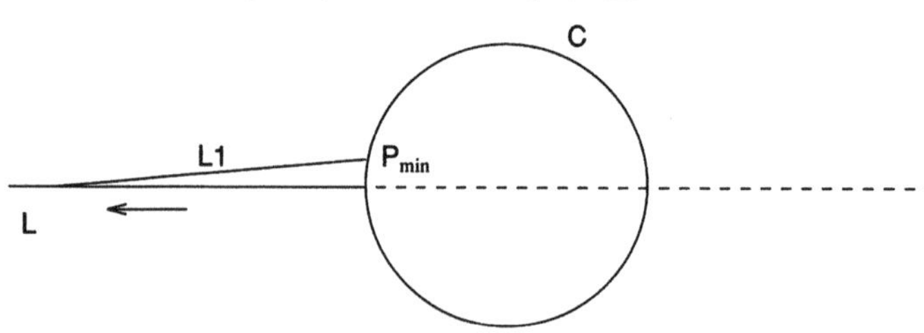

We mention a few interesting observations about our case study:

- A designer can deal with incomplete information when working on a design. He may even finish a design without complete information to imply a unique solution. This capability is usually missing in other constraint-based systems. The systems based on the algebraic interpretation of constraints cannot solve this problem because some underlying constraints are only implicitly specified. For example, the following two constraints are necessary to imply a valid embedding:

$l_1 \geq d+r$
$l_1 \leq d-r+l$

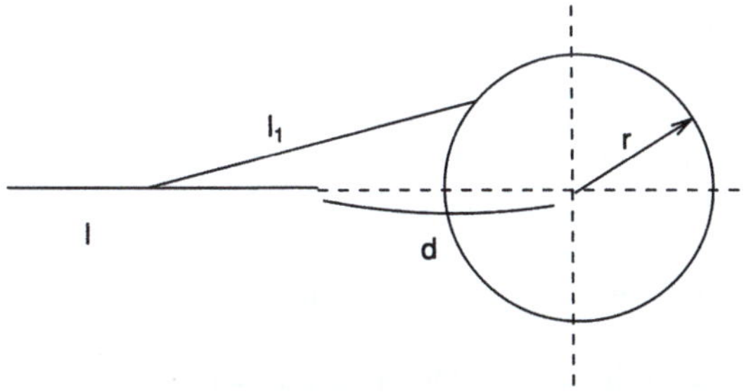

- Qualitative constraints and those involving inequalities can be more easily satisfied if a designer can interact with the system. For example, pick a center for C on the extension of L, place a circle above a line, put a circle inside a box, etc.
- Reasoning about spatial relationships symbolically simplifies the computations and avoids problems with numerical methods.

8. Discussions

Some of the important characteristics of our constraint management system are summarized below.

- Two Tiered Model

  Incompleteness of information is inherent in design. Frequently, not enough information is available at intermediate stages of a design to derive a system of algebraic equations with unique solutions. Defaults and arbitrary fabrication of additional equations are commonly used to derive a unique solution in many systems. Coercing completeness in this manner has the drawback that once such additional constraints are added, they are often treated indistinguishably from the 'real' constraints.

  In OTP, there is a clear separation between the intended constraints and the additional information necessary to render one of the possibly many solutions that satisfy those constraints. The system uses two models in parallel: a constraint model and a sketch model. The constraint model is almost always incomplete, reflecting only those constraints explicitly required by a designer. The sketch model is always complete, reflecting in the form of a concrete sketch, both the implications of the constraint model and all additional information necessary to drive a solution.

  In most systems, there is a tight correlation between the sketch and the set of constraints. We believe that while sketches are important tools for conveying relevant design information, not everything contained in a sketch should be taken seriously. Typically, most metric information and some topological relationships implied by sketches prove to be either wrong or irrelevant as a design evolves. Our design system assumes a looser binding between constraints and sketches than most other constraint based systems: sketches and constraints may be modified directly and independently. Changing a sketch or its underlying set of constraints may or may not entail a change to the other.

- Incremental Satisfaction

  Refinement of ideas during the course of a design often results in incremental addition of constraints. It is conceptually simpler to ignore existing solutions, start with the modified set of constraints, and try to find a solution to satisfy them. This approach suffers from three drawbacks. First, there is extra work involved in resatisfying the bulk of the constraints that remain intact after each incremental modification. Second, due to instability of some numerical methods and limitations of equation solving techniques, it may not be possible to solve the old equations in the presense of the new ones. Third, the solution found for the new set of constraints may not be related to the one for the old set of constraints in an intuitively meaningful way.

  In OTP, incremental addition of a constraint results in local modifications to the existing solution (sketch model). This can be more stable and computationally less expensive compared to constraint satisfaction systems based on algebraic techniques. The advantages become significant especially in the context of interactive CAD, where incremental growth and editing of a design's underlying network of constraints is predominant.

- Geometric Reasoning

  Most constraint systems are primarily based on the algebraic interpretation of geometric constraints, and lack the ability to reason about the abstract properties of constraints and design objects. In OTP, the higher-level semantics of geometric constraints are used at a planning stage to derive the local changes necessary for satisfaction of a constraint.

  The semantics of a satisfied constraint is expressed through the degrees of freedom that it leaves for its participating objects. The degrees of freedom of an object under a constraint are represented as a set of constraint-preserving operations. A new constraint is accommodated by a network of satisfied constraints through propagating the sequence of operations associated with the new constraint to the appropriate previously-satisfied constraints. Similarly, the semantics of satisfying a constraint is also expressed through a set of operations.

  Constraint satisfaction in OTP uses symbolic reasoning to plan a sequence of operations that transforms an unsatisfied set of constraints to a satisfied set of constraints. It can use the higher-level semantics of geometric relations to avoid unnecessary computation. For example, knowing that tangency between a line and a circle is invariant under rotation around the center of the circle, an OTP system can avoid using the algebraic interpretation of tangency in appropriate contexts.

  In systems based on the algebraic interpretations of constraints, sometimes the transitive closure of equations involving some given variables is used to localize solutions. This technique is purely syntactical and disregards the semantics of the context of constraints. As a worst case scenario, this technique may end up solving the whole set of constraint equations, whereas our approach can localize the changes (and equation solving) as shown in the following example.

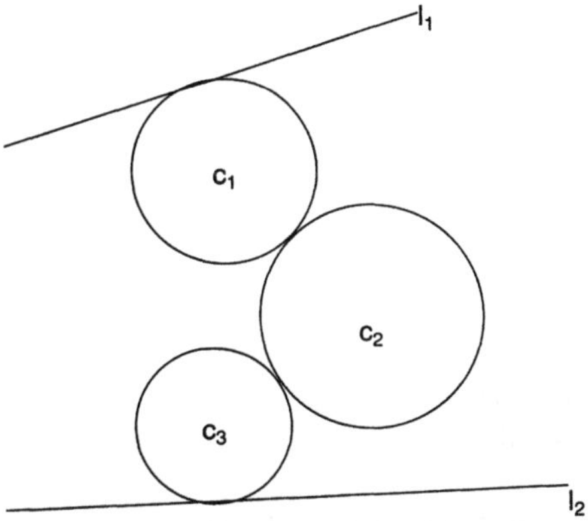

Fig. 9. Two constraint lines and three constraint circles

In the example of Fig. 9, the constraints are l_1 tangent c_1, c_1 tangent c_2 $c2_1$ tangent c_3 and l_2 tangent c_3. To impose the constraint l_1 parallel l_2, the algebraic approach will solve the equations corresponding to the five constraints. Our approach will consider only a proper rotation of l_1 around the center of c_1 such that l_1 is parallel to l_2. The computation involves l_1 and l_2 only, and the only change is that of the slope of l_1.

- Qualitative Constraints

  The operational transformation approach used in OTP provides a convenient means for dealing with qualitative constraints. It is sometimes difficult to represent a constraint analytically, in terms of algebraic equations (e.g., symmetries). Furthermore, constraints involving inequalities and higher degree polynomials are sometimes difficult to solve. On the other hand, it is often possible to devise procedures for satisfying and maintaining a constraint using an appropriate set of primitive operations.

9. Conclusion

We presented a constraint-based design system based on the Operational Transformation Planning (OTP) approach. In OTP a constraint is satisfied by devising a sequence of operations that transform a given configuration to one where the constraint holds. The construction techniques of Euclid in his proofs of elementary geometry theorems are good examples of the operational approach to constraint satisfaction. The OTP approach is quite suitable for interactive design because of its incremental satisfaction of constraints and preference for local modifications. OTP tries to avoid transforming constraints into systems of algebraic equations, when possible. Thus, global computation, e.g. solving systems of equations every time a dimensional value is changed, is replaced by local computations incorporated into functional procedures. Furthermore, through geometric reasoning, OTP can avoid some geometric inconsistency problems. In our system, abstract information and embedding information of a design are separated and dealt with at different levels, The effects of a change to a design are computed locally through a sequence of well-defined operations. Therefore, OTP systems can be semantically more robust and computationally less expressive than systems based on the algebraic approach. Especially, OTP systems can deal with a wider range of geometric constraints, such as qualitative constraints which are difficult to incorporate into the existing CAD systems.

References

1. F. Arbab and J.M. Wing, ''Geometric Reasoning: A New Paradigm for Processing Geometric Information,'' in *Design Theory for CAD, Proceedings of IFIP W. G. 5.2 Working Conference 1985 (Tokyo)*, ed. H. Yoshikawa, pp. 107-121, North-Holland, Amsterdam, 1986.
2. F. Arbab and B. Wang, ''Reasoning about Geometric Constraints,'' in *Preprints of the 2nd IFIP 5.2 Workshop on Intelligent CAD*, Cambridge, UK, 1988.
3. F. Arbab, ''Preliminary Report on Oar: a System for Objects And Reasoning,'' Technical Report, Computer Science Department, University of Southern California, 1988.
4. F. Arbab, ''Examples of Geometric Reasoning in Oar,'' in *Intelligent CAD Systems II: Implementational Issues*, ed. V. Akman, P.J.W. ten Hagen and P.J. Veerkamp, pp. 32-57, Springer-Verlag, Berlin, 1988.
5. A. Borning, ''The Programming Language Aspects of Thinglab, A Constraint-Oriented Simulation Laboratory,'' *ACM Transactions on Programming Languages and Systems*, vol. 3, no. 4, pp. 353-387, 1981.
6. B. Bruderlin, ''Constructing Three-Dimensional Geometric Objects Defined by Constraints,'' in *Proceedings of the 1986 Workshop on Interactive 3D Graphics*, pp. 111-130, ACM Press, 1986.
7. J. Gosling, ''Algebraic Constraints,'' CMU-CS-83-132, Department of Computer Science, Carnegie-Mellon University, 1983. PhD Dissertation
8. R.C. Hillyard and I.C. Braid, ''Analysis of Dimensions and Tolerances in Computer-aided Mechanical Design,'' *Computer-Aided Design*, vol. 10, no. 3, Butterworths, 1978.
9. F. Kimura, et al., ''Variational Geometry based on Logical Constraints and its Applications to Product Modelling,'' in *Proceedings of CIRP '87*, 1987.
10. F. Kimura, H. Suzuki, and L. Wingard, ''A Uniform Approach to Dimensioning and Tolerancing in Product Modelling,'' in *Proceedings of CAPE '86, Second International Conference on Computer Applications in Production and Engineering*, pp. 165-171, 1986.
11. H. Lieberman, ''Using Prototypical Objects to Implement Shared Behavior in Object-Oriented Systems,'' *special issue of ACM SIGPLAN Notices*, vol. 21, no. 11, pp. 214-223, 1986.
12. H. Lieberman, ''Delegation and Inheritance: Two Mechanisms for Sharing Knowledge in Object Oriented Systems,'' in *3eme Journees d'Etudes Langages Orientes Objets*, ed. J. Bezivin and P. Cointe, AFCET, Paris, France, 1986.

13. R.A. Light and D.C. Gossard, ''Modification of Geometric Models through Variational Geometry,'' *Computer-Aided Design*, vol. 14, no. 4, Butterworths, 1982.

14. R.A. Light, ''Variational Geometry: Modification of Part Geometry by Changing Dimensional Values,'' in *Proceedings of Conference on CAD/CAM Technology in Mechanical Engineering*, MIT, 1982.

15. V.C. Lin, D.C. Gossard, and R.A. Light, ''Variational Geometry in Computer-Aided Design,'' *Computer Graphics*, vol. 15, no. 3, Association for Computing Machinery, 1981.

16. J.R. Rossignac, ''Constraints in Constructive Solid Geometry,'' in *Proceedings of the 1986 Workshop on Interactive 3D Graphics*, pp. 93-110, ACM Press, 1986.

17. J.R. Rossignac, P. Borrel, and L.R. Nackman, ''Interactive Design with Sequences of Parameterized Transformations,'' in *Intelligent CAD Systems II: Implementational Issues*, ed. V. Akman, P.J.W. ten Hagen, and P.J. Veerkamp, pp. 93-125, Springer-Verlag, Berlin, 1989.

18. D. Serrano and D. Gossard, ''Constraint Management in MCAE,'' in *Artificial Intelligence in Engineering: Design*, ed. J.S. Gero, Computational Mechanics Publications, 1988.

19. G.L. Steele, Jr., ''The Definition and Implementation of a Computer Programming Language Based on Constraints,'' AI-TR-595; MIT Artificial Intelligence Laboratory, 1980. PhD Dissertation

20. G.J. Sussman and G.L. Steele Jr., ''CONSTRAINTS - A Language for Expressing Almost-Hierarchical Descriptions,'' *Artificial Intelligence*, no. 14, pp. 1-39, 1980.

21. I. Sutherland, ''Sketchpad, A Man-Machine Graphical Communication System,'' Phd thesis, MIT, 1963.

15

Discussion Session: Geometric Reasoning

Paul ten Hagen

1. Papers

The three papers in this session use objects and constraints to represent geometric problems. The first paper, presented by Rajiv Desai studies the geometric reasoning module as part of a complete system for geometric analysis. A user asks predefined tasks to be performed, where each task is translated in a combination of geometric modelling, reasoning and motion planning activities.

The second paper, presented by Can Baykan, uses objects and constraints in 2D-space planning. It concentrates on the search strategy over the state space of partial solutions. The user interface allows for closely observing the search process and for providing directions during the search.

The third paper, presented by Farhad Arbab proposed a constraint management system to experiment constraint-based design for mechanical engineering applications. The system supports geometric objects with mutual relations represented as constraints. For each constraint operations are defined on the objects to satisfy it and to retain it when additional operations and/or constraints are imposed.

2. General Discussion

On the paper by Desai the issue was raised about the generality of a geometric reasoner in a general purpose Geometry Analysis system. Especially the question of whether this generality will be defeated by poor efficiency was discussed. No conclusion was reached about this point, although several people expressed their agreement with the approach taken.

Baykan's paper was rather complementary to the previous one. It tried to find the best search strategy for a rather specific space planning problem. 'Best' meaning efficient, flexible, and user friendly. It could be seen that the example problem could be generalized to a large class of optimal-layout problems. The analysis of the reason why the preferred (right) method was the best one suggested that it would remain the best for more general cases.

Arbab's paper provided a conceptually consistent approach to geometric reasoning including a system architecture which is a realization of the computational model for geometric reasoning. In particular, it convincingly showed the need for planning a constraint satisfaction process before actually starting it. It was extensively discussed how general geometric knowledge can be incorporated next to problem specific geometric knowledge. The former being needed for constraint satisfaction planning.

User Interface

16

POINTER: Picture Oriented INTERaction

A Programme for ICAD/HCI Research

Aart Bijl

EdCAAD, Univ. of Edinburgh
20 Chambers Street
Edinburgh EH1 1JZ, UK

Abstract: This paper outlines a general strategy for ICAD/HCI, providing a context for collaborative research projects. An instrumentalist approach to computer systems is advocated, together with a quasi-linguistic treatment of graphics, leading to systems by which users can construct and modify expressions of their own knowledge.
Keywords: Knowledge, CAD, Graphics, Representation, User-Interfaces, Design

1. Introduction

Long experience of CAD and, in more recent years, the application of logic to design has led me to a key observation about human/computer interaction (HCI). This is that satisfactory HCI cannot be achieved by interposing an interface between a user (a designer) and some otherwise separate computer system – HCI requires user control of the whole of a computer system when it is applied to some design task. This interest in possible user interactions, and in what users need to know in order to interact with a system, entails a fundamental reappraisal of the nature and role of systems.

A user has to know how to exploit the surface behaviour (the functionality) of a system on design tasks. Here we appear to be faced with a choice between two strategies which can be distinguished by referring to two extreme hypothetical positions. Either a system is thought of as possessing a lot of domain knowledge plus procedures for finding answers to specific instances of design task – the view of systems which sees them as possessing human intelligence. Or a system is thought of as possessing abilities to perform domain independent operations on user input – the view of systems which sees them as useful for purposes of realising expressions of users' own intelligence. The former is exemplified by expert systems, and the latter by word processors.

If we look hard at the difference between these extremes, we will find that a precise definition of the difference becomes elusive. In the case of artificial human intelligence, the functionality of the system might be embedded (or hidden) in expressions from the system which appear close to a user's own domain knowledge. In the case of domain independent systems, their functionality might appear obviously disjoint from a user's domain knowledge. In both cases the user has to employ the given system functionality in order to realise some result. The problem presented by the former case is that the embedded functionality of the system is likely to 'get in the way' of designers.

In this paper I will advocate the latter extreme position, focusing on the usefulness of domain independent systems. I do so not out of strong conviction that this offers the right and only strategy for CAD, but because it is an interesting and potentially fruitful position which tends to be neglected by enthusiasts of intelligent systems.

The strategy generally adopted for ICAD (intelligent computer aided design) assumes the possibility of achieving correspondence between computational models of domain knowledge and a user's own knowledge, and that this correspondence can be valid even if it is not visible to designers. I will try to show how EdCAAD's MOLE system (Modelling Objects with Logic Expressions [17])[1] is based on a different strategy. This accepts that we cannot know to what extent any prior model of domain knowledge actually is the same as a designer's own knowledge. Instead, we see domain models as occurring in readings of expressions at the interface between a computer and a designer, and we see these readings as remaining the responsibility of the user. The role of MOLE is to support production and modification of expressions that get passed between people and, to serve this purpose, the logic which the system offers must remain visible and accessible to users. This strategy may provide the only way out of the prescriptive imperative of orthodox ICAD systems on design practices.

This is a major objective and this paper outlines a programme for collaborative research. It seeks to stimulate responses from within the research community, leading to collaboration that will be motivated by interests in fundamental research.

2. Artefactual Systems

We can start with a simplistic view of systems as either:

1. rational descriptions of behaviours of naturally occurring things, where we do not know what the parts of those things actually are and we do not have control over their behaviour – this understanding comes close to the notion of theories which rest on hypotheses about behaviours of naturally occurring phenomena;
2. artefacts that we make with the intent that they should exhibit certain desired behaviours, responding to inputs and producing results – this understanding comes close to the notion of machines, as indicated in Fig. 1, and it is this understanding which is meant by artefactual systems.

---

[1] MOLE has been developed with support from the Building Sub-Committee of the Environment Committee of the UK Science and Engineering Research Council, and has been described at the two previous ICAD Workshops.

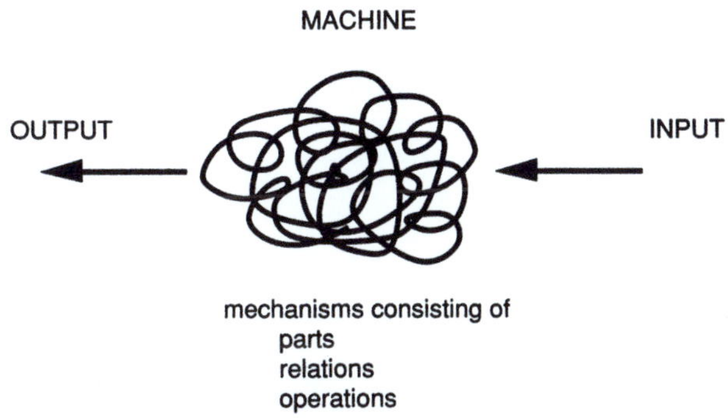

Fig. 1. The concept of a machine

These two positions can be viewed as being closely related. Both employ the concept of behaviour as something which is seen to happen, without necessarily knowing all the operations which might occur within the behaving thing. The distinction between naturally occurring phenomena and artefacts made by people can become obscure. Realisations of expressions of rational descriptions will be artefacts (as explained in my discussion of language [3]).

Nevertheless, the distinction between those two positions does have some purpose. Expressions of rational descriptions traditionally are articulated and evaluated only in the minds of persons. It is the behaviours of these minds which is depicted in expressions, and those expressions are used to support interactions between minds. If we expect expressions to be acted upon by artefactual systems (not minds), and if we expect those actions to be conditioned by logic, then we are moving into the realm of machines and computers. We can view logic as serving to specify operations which have to be executed within computers, so that they will exhibit logical behaviour.

When people use machines they have to know the range of possible behaviours of their machines, without necessarily knowing the internal operations that are required for such behaviours. Furthermore, the machines have to 'know' the range of possible inputs from users which can invoke operations that will result in behaviours. This need-to-know is indicated in Fig. 2. However, a machine does not need to know why a user is calling for a certain behaviour. For example, a car needs to sense its driver's use of its controls, but does not need to know why the driver wants to go to any particular destination. Similarly, a typewriter needs to sense its user's key hits, without knowing why the typist is composing a particular piece of text.

This relationship between machines and users persists when machines are computers, in the familiar example of word processors indicated in Fig. 3. The computer has to be able to sense key hits that will invoke its internal operations, so that it will exhibit behaviour in realising arrangements of character strings. Typically, a computerised word processor offers increased functionality, when compared with a typewriter, by being able to insert, delete, move and replace pieces of text, and by controlling font style and text formatting. This increased functionality does not depend on the machine knowing why a user wants to compose a particular piece of text.

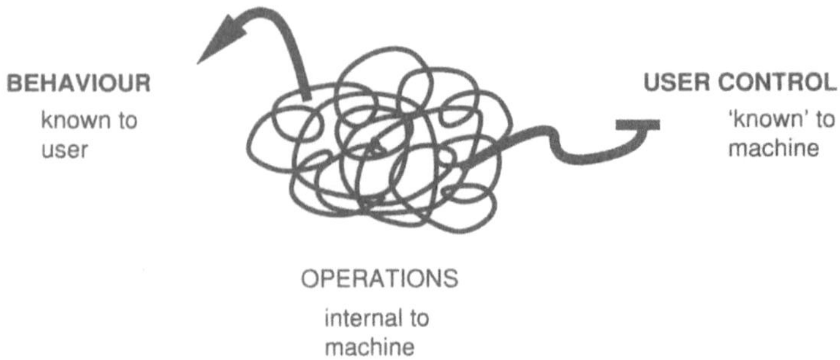

Fig. 2. Behaviour of machines

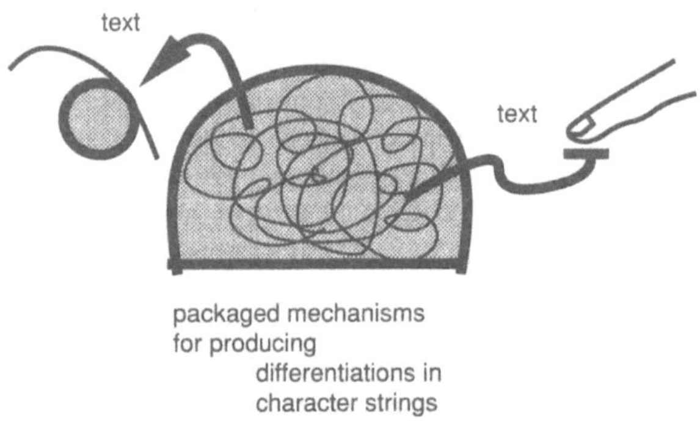

Fig. 3. A machine for processing expressions

This understanding of machines can be applied generally to most (if not all) artefactual systems. It has to apply if we expect to make machines in large numbers, which will be useful to many people. Cars are successful (at least in terms of their existence in large numbers) because they are effective in moving people without determining where they must go. Word processors are successful because they are effective in realising and modifying character strings without determining what words they must depict.

The position which I am advocating in this paper is that we ought to hold on to this understanding of machines even when we become ambitious in our expectations of future computer systems. In recent years we have rightly congratulated ourselves on our progression along the 'how-to-what' spectrum, affecting users' access to computational processes. But this advance can be viewed simply as computers catching up with other more established machines which offer users better control over their behaviours. Extending the spectrum beyond 'what' to include 'why' takes us beyond machines and into human intelligence.

Having stated this position, I recognise that it is not clear cut. It is not clear how we can arrive at a precise and durable distinction between 'what' and 'why', and this distinction probably rests on changing social norms. There is much scope for research here, especially in fields of design, including mechanical engineering. For the moment we can say that if a 'what' provokes a 'why' in a user domain, then it should not be embedded (hidden) within a computer system.

This position has conditioned EdCAAD's development of the MOLE system, aimed at extending functionality beyond word processors. Increased functionality is obtained by including logical connectives for relating pieces of user expressions in the form of either text or pictures. Our intention is that control of the effects of logical connectives should remain visible and accessible to users. We are trying to hold on to the idea that the behaviour of the system should not depend on prior modelling of users' domain knowledge.

2.1. An instrumentalist approach to computers

The strategy adopted for MOLE can be regarded as an instrumentalist approach to computers. Quite simply, an instrumentalist approach means that we look upon computers as instruments for doing things, without the implication that they model people or the things that people do. Computers can be regarded in the same way as we regard other tools, such as mechanical tools which are used because they enable us to do things. We don't think of those tools as modelling ourselves – they just do things that we find useful. Computers are no different. The ramifications of this position run deep, and only a few consequences will be considered here.

The thing that a computer does is perform operations on expressions, including the execution of mappings between different forms of expression, and the application of formal systems (such as arithmetic) to expressions. In all cases, these are expressions from persons and we can regard them as representations of what the persons know – but a computer does not know that they are representations.

By executing complex sequences of operations on many expressions, a computer can be made to look like it is behaving as a person. Thus we are tempted into a common confusion over the distinction between a representation and whatever it represents, especially if it is thought to represent human knowledge. We are tempted to try to make computers be more like persons, or make them do things more like persons do them, and make computers that can understand the behaviour of persons. We find ourselves embarking on a major enterprise to devise computational models of the world of people, and of persons in this world. However, even sophisticated computers remain just machines for executing operations on expressions. We should not reduce the impact of this fact by claiming that humans are no different [3, 4] – perhaps we cannot know (in an overt sense) what expressions represent within persons, and we ought to accommodate the possibility that expressions and whatever generates them from within persons are different.

An instrumentalist approach to computer systems accepts them as tools, albeit highly sophisticated tools, which persons can use to construct their own expressions for purposes of communicating with other persons. This includes the possibility of constructing expressions that have the effect of modifying other expressions, and the possibility of persons gaining access to and using each other's expressions. A computer can support such activity, operating on the structures of expressions without it knowing what the expressions might mean in the minds of users – just as more orthodox machines can operate on properties of other artefacts, without knowing why we might want them to do so.

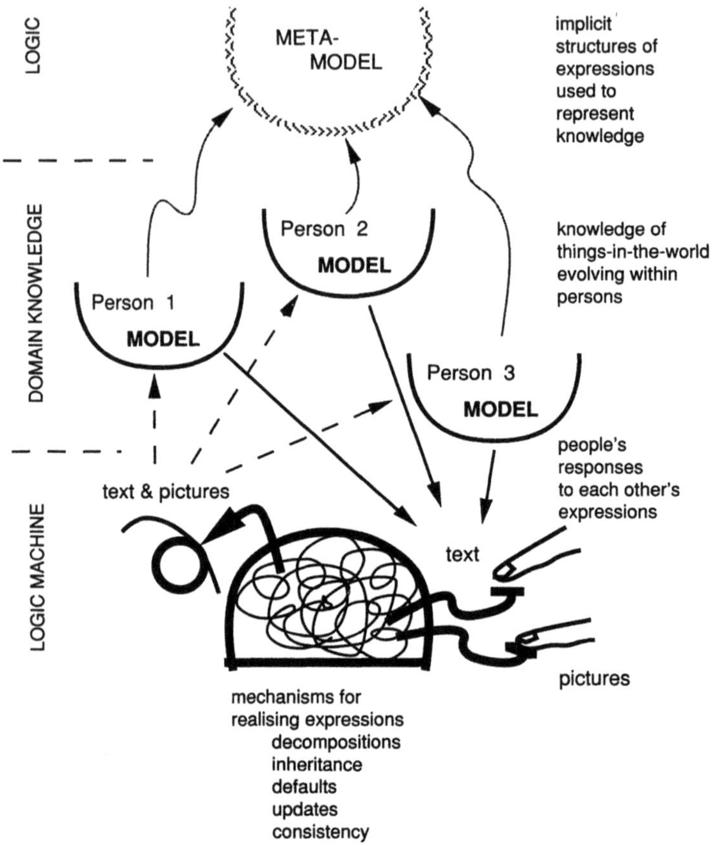

The instrumentalist approach to HCI rests on a distinction between what might be happening in a person and what can happen in a machine, and the use of meta-models to decide the functionality of a machine. Domain knowledge and abstract models for such knowledge reside within persons. They can use machines to realise and modify expressions of their knowledge, and they interact with each other by means of expressions to modify states of knowledge within themselves. The instrumentalist approach sees the role of meta-models as determining domain-free logical structures for expressions, which can be supported by machine operations and used by people.

Fig. 4. Human/computer interaction

The strong implication here is that persons using computers to construct expressions have to know the tools which they are using, and have to retain responsibility for their expressions. Persons have to know the behaviour of computers for purposes of constructing expressions, without necessarily knowing a computer's inner operations which support that behaviour. Here we come to the essential focus of human/computer interaction, the study of inner operations of computer systems which can support behaviours appropriate to people's construction of expressions. This focus is illustrated in Fig. 4.

Fundamentally, this position is compatible with those of cautious AI advocates, but it concentrates our attention on how the role of computers can or should be viewed by users. Computers are seen here as tools offering logical inferencing mechanisms which people can use to express their own models of their domain knowledge – these inferencing mechanisms can be similarly applicable to pictorial and written forms of expression.

2.2. The example of MOLE

MOLE is an implementation of the kind/slot/filler convention based on a notion of things (any kind of concrete or abstract thing) being described by further things – without prior conditions on what those things can be. In keeping with our instrumentalist approach, we are interested in the essential K/s/F structure, rather than the variants that have been developed for particular purposes, such as frames and scripts. Our structure is indicated in Fig. 5. It employs kinds as names for entities, slots as names for parts of those entities, and fillers as expressions that point to further kinds as parts which which fill slots. The resulting parts hierarchies can be interlinked through inheritance, to form networks. In the general case of this structure, it can be used to compose unbounded descriptions of parts, and inheritance allows changes within descriptions of parts to propagate changes within other parts.

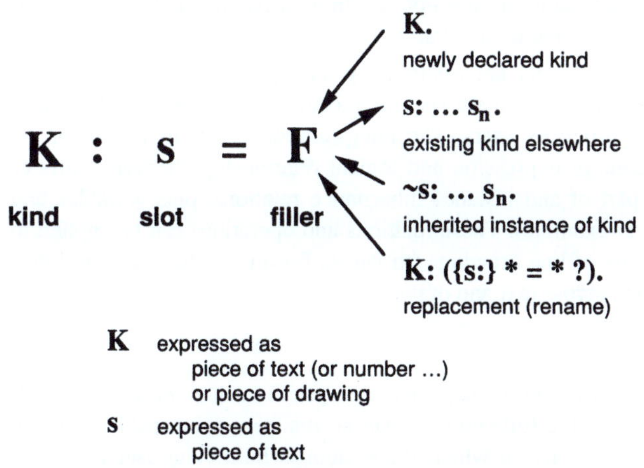

In this structure kinds name concrete or abstract things, and they can be expressed in the form of words, numbers or other specialised symbolic notations, and in the form of drawings. Slots name parts that describe kinds. Fillers are expressions which refer to further kinds as parts which fill slots, and they can include functions and constraints. These expressions get evaluated by the system, to effect changes or additions to descriptions of kinds. A filler can be: a kind, which evaluates to itself; a path name or indirection consisting of a succession of slots, which evaluates to the kind at the end of the path; an instruction to make a new instance of the kind at the end of the path; an instruction to replace a description of one kind with the description of another kind; plus further instructions that will become apparent in the following discussion.

Fig. 5. MOLE's **K/s/F structure**

In the K/s/F convention, descriptions are collections of slots to a kind, naming the parts that describe that kind, successively down parts hierarchies and across inheritance networks. So descriptions are logically related pieces of user expressions. All this forms a structure which can be used to represent domain knowledge in the form of user declared expressions. To use a system based on this structure, users need to satisfy themselves that they have appropriate interpretations for the structure – they have to know the system.

In principle, this sort of system is able to receive unanticipated descriptions of design objects from users, in the form of newly declared kinds, and add them to its database. These descriptions can include conditions under which previous or future descriptions will be changed, and can include descriptions of the changes. In effect, the system can allow users to construct a database (or knowledge base, but without the implication that its operations match the way that users think) as well as interrogate a current state of the database.

Users' domain models in the form of drawings and text are represented within MOLE, in a logical form, and it is conceivable that different users will invoke each other's models from within MOLE. The difference in the schema presented by MOLE, compared with the more usual approach to ICAD, is that domain models serve as models only when users read them from realisations in the form of drawings and text. MOLE itself carries responsibility only for the consistency and coherence of its representation, and not for the consistency and coherence of domain knowledge. It is the users who are the modellers.

In this schema, abstract domain models are retained by users and the system provides support for realisations of these models. No domain specific models are held within MOLE in a manner that cannot be seen and modified by users. The interpretation functions on which the system is built, deals only with forms of expression and logical relationships between parts of expressions. The system includes part of and instance inheritance relations, plus variables and conditions, and evaluations (see later example). These relations and operations can be applied to parts of character strings, and to parts of line drawings (to values for angles, lengths, and kinds of attachments), and they can be used across text and drawings.

2.3. User interfaces

The problem which now emerges is how can a designer use a system such as MOLE – what is expected of the user? From experience, the following answer seems to be inescapable: the user has to be familiar with the logical relations on which the system is built. The user has to be familiar with logic in the abstract, in order to use the machine's implementation of logic to produce and interact with expressions, and to control the accumulation of expressions that is represented and stored within the system. The user has to know how the system's use of logical relations affects the consistency and coherence of the machine's representation of his or her design knowledge.

For practising designers, this is an awesome and unfamiliar responsibility. It entails a mode of thought that appears to be foreign to the activity of designing. In effect, we are asking designers to program their computers.

It is usually argued that this observation can be moderated in a number of ways. In programming their own machines, designers can expect to reuse their programs, recalling old programs, so that they do not always start with an 'empty' machine. They can reuse programs that make the machine exhibit desired behaviours in producing and modifying expressions. Different designers might share each other's programs. However, while all this might be true, this speculation does not remove the fundamental requirement that the designer has to know and become familiar with the logic on which the system is built – otherwise the designer will not be able to modify existing programs to meet changing demands that come from the world of design.

More commonly, it is argued that other people should program computers for designers. Here we find the old distinctions between system and application programs, and computer programmers and end-users – applications depend on users' knowledge being programmed in some prior way, within computers. The objection to this strategy is that it leads to prescriptive systems which compel users to work within the bounds set by application programmers

As a variant of this orthodox strategy, we have 'user friendly' systems – exemplified by the Macintosh's standard window/icon/menu interface, used in the same way by different application programs. Friendliness seems to refer to the surface behaviour of a system, evident in a graphical form which is attractive to users and constant across applications. This works for regular 'simple' applications but causes problems when users have particular needs, when they find themselves in contention with detailed decisions taken by application programmers. The friendliness of the interface then obscures the logic of the system from the view of the user, and results in highly prescriptive programs.

Lastly, we should note that even in the case of these friendly systems, users have to program their knowledge into these systems – they have to work within the limits of the procedures that constitute computer applications.

All these strategies do not remove the fundamental requirement that a designer has to know the logic of the system. It is said that people ought to be logical anyway and, therefore, for sensible people, this requirement ought not to be too onerous. Such a position assumes an equivalence between computational logic and ordinary human logic – an assumption that is difficult to sustain. Instead, we might see the problem of designing an ICAD interface as one of reconciling the difference between computational logic and the separate logic of users. This problem is profound if we accept that we cannot achieve a prior formal understanding of the logic of users, as is the case when users are designers. Research then needs to focus on the problem of how to make the logic of a system visible, accessible and controllable by users, so that designers can decide how to use the system to express their own design knowledge.

3. POINTER

POINTER is a proposed framework for a collaborative programme of research, arising out of experience of MOLE and earlier CAD systems, and more recent work on combined NL and graphical interaction with a computer's knowledge base. This programme is aimed at a mode of human/computer interaction, HCI, which will allow persons to interact with computational processes by means of pictures. The focus of this interest is on the construction of graphical expressions with interpretations to other (symbolic) representations. The purpose is to achieve system generated responses which modify graphical expressions, in a manner that will correspond with user expectations. For this purpose, we propose a quasi-linguistic approach to graphics, covering:

- meaning of graphical compositions in context of use;
- representation of that meaning in computational symbolic logic;
- use of an interactive interface to extend and change meaning.

This emphasis on pictures used in conjunction with words is crucial to the development of an acceptable HCI system for ICAD.

3.1. Market

The potential market for POINTER developments is very large, including the kind of computer users who currently use Macintoshes and PCs with graphics. These developments will be useful in design fields where graphics is used to express and manipulate information, making CAD applications more accessible and controllable by designers. More generally, these developments will be useful to managers and policy makers in commerce and industry, who want to be able to control their views of information held in computers and who find pictorial presentations of complex information more understandable.

POINTER developments, extending beyond the paradigms of Mac HyperCard and object-oriented programming, can be expected to contribute to the next advance in popular and friendly user interfaces.

3.2. Background

Present HCI systems that use pictures, as exemplified by Macintosh icons, offer graphic primitives which serve as commands or 'canned phrases'. Users cannot compose these primitives into newly articulated graphical expressions with interpretations. These systems also allow mixing of bit map images, but they 'know' of no further structure of images which could support interpretations to other representations of things that images depict. These limitations restrict users' graphical access to computational processes which might otherwise support representations of their own knowledge.

Early CAD systems [2, 6] offered some facilities for constructing and interpreting graphical expressions, but these were highly domain specific. Attempts to achieve generality [18] have demonstrated potential, incorporating user models and semantic integration with the displayed images, but they have concentrated on graphical presentation. More recently, in the field of geometry modelling, there has been renewed recognition of the need for user control of concepts which relate graphical expressions to associated knowledge of materials, techniques and client specifications [10, 12].

A linguistic approach to graphics has been adopted for a large ESPRIT project, (P393) 'ACORD', which includes EdCAAD and the Edinburgh Centre for Cognitive Science among its partners [7]. Results are promising [9, 11, 16]. That project covers formal treatments of predefined graphic objects which are used as system primitives with certain variable properties and relations. Now the aim is to progress towards user defined graphical compositions with user defined meanings.

More generally, an appreciation of graphics as a communication medium is relatively undeveloped in the fields of computer science, artificial intelligence, and cognitive science. Few cognitive psychologists have looked into the mechanisms of picture production and associated interpretation [14]. How pictures can be used to convey varied kinds of information remains an important question, and new research needs to contribute to and stimulate interest in this area.

4. Proposed system strategy

Our proposed strategy for HCI, as indicated in Fig. 6, is based on the observation that all expressions at the interface between a user and a computer, on a screen, are in graphical form. Graphics here comprises 2-D spatial arrangements of lines or other graphic objects, and includes text characters. A representation for graphics can therefore comprise mappings to pictorial objects defined in terms of their syntactic categories (i.e. possible readings of graphical compositions), and to textual objects defined in terms of similar categories for character strings. These objects can be 'known' in lexica of picture parts and words so that they and their associated relations can be used by further computational processes.

This grouping of text under graphics may appear strange – but we are interested in operating on spatial relations in compositions of pictorial and textual objects. We want to include the possibility of attaching significance to variations in depictions of words.

Linguistic treatment of pictures rests on lexical definitions for parts of pictures, which support parsing and semantic representations for pictorial compositions. The separate functionalities of pictures and words can then be combined in a system's composite representations of screen images. Functional models represent tasks that can be applied to interpretations from graphical images, such as generating 3-D spatial projections from 2-D pictures. They can receive and act upon pictorial and textual objects, and pass results to the system's knowledge base, as users' domain specific models which can be recalled in the form of graphical images on the screen.

Our focus on expressions sees them as representations of human knowledge, covering a spectrum from general knowledge of treatments of expressions as language, to domain knowledge described through use of language. A system that represents treatments of expressions, built on lexical entries (like Smalltalk† objects, but oriented toward linguistic behaviour) with semantics known to users, can be used to construct and modify expressions of domain knowledge, the kind of expressions that are passed between people. The functionality of the system will then consist of its application of inferencing mechanisms to uses of lexical entries and, consequently, to users' graphical expressions of their domain knowledge. We see here an important distinction between the system's use of its internal resolution facilities which condition its representations of user expressions, and users' use of their knowledge about lexical entries when constructing expressions – the latter gives users control over the behaviour of the system.

The system has to make use of differentiations which it can 'read' in its representation for graphical images. This applies to differentiations in arrangements of lines depicting picture parts

† Smalltalk-80 is a registered trademark of Xerox Corporation.

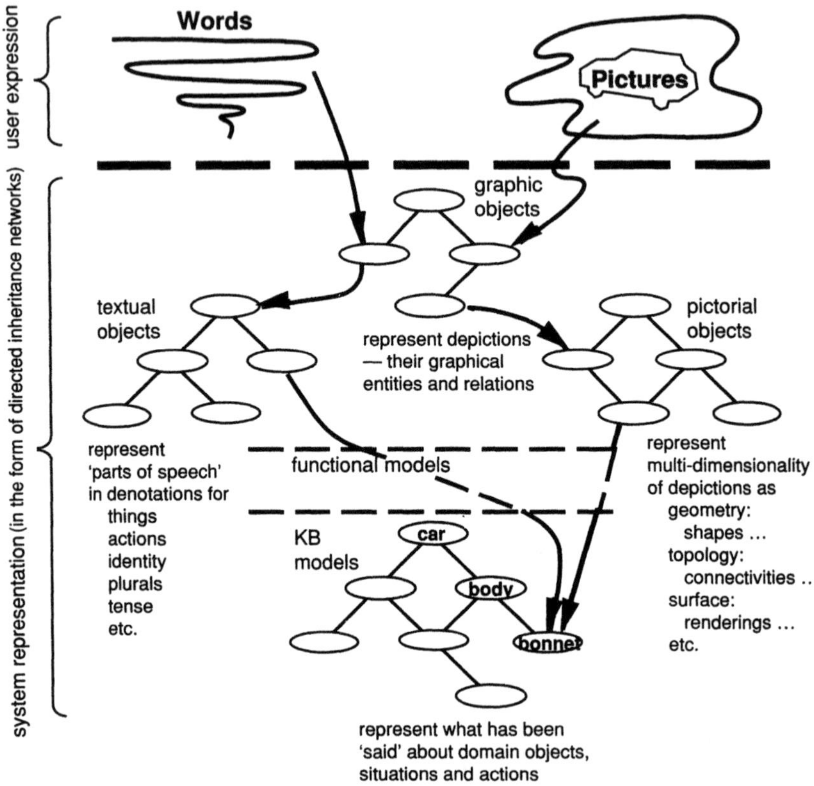

This is a tentative picture of the strategy outlined in this paper – and it is an example of a graphical expression which combines pictorial and textual objects.

Fig. 6. Combining words and pictures

and to differentiations in strings of text characters. In the case of pictures, differentiations in images can be mapped to pictorial objects that might be defined in terms of: classes of boundary conditions to 2-D spaces, such as lines (with angle and length) and polygons (with area I); connectivities between such objects, as topologies; surface treatments of objects, as in colour and texture rendering.

Such dimensionalities need to be calibrated against (arbitrary but agreed) units of measure, for purposes of identifying values which can be passed from the image through the lexica and on to functional models.

A user will normally need to be aware of both pictorial objects and textual objects, by means of text and other depictions on the screen. The user will have to know enough about these objects in order to use them to construct expressions of her or his own domain knowledge.

The system's symbolic representation for uninterpreted images will not normally need to be known by users. However, we can expect that the structure of a representation, corresponding to the extents and connectivities of graphic entities in an image, will be evident in the depiction on the screen.

5. Research Goals

The aims of POINTER can be supported by research on separate fronts, on separate aspects of this system strategy. Advances on each front can contribute to the success of a composite system for effective HCI. However, an advantage of the strategy which has been outlined is that no one system component should be critical to the realisation of other components – the strategy allows for piecemeal advances to be assembled into ever better systems. This possibility rests on the necessary simplicity of the system's control mechanisms, which must not be too 'clever' (i.e. must not presume too much about the intentions of particular users). The aim is to allow users to exercise control over expressions and the system's corresponding representations of their own domain knowledge.

5.1. Research contributions

Research contributions can be focused on the following topics, with the intention that advances should have general relevance to future developments of commercial systems.

5.1.1. Image representation. Definition of basic graphic entities, properties and relations from which any images can be constructed – serving as a graphical modelling environment for graphical productions. The research problem here is to distinguish between the graphical modelling environment and pictorial objects that can be modelled. To achieve generality, we have to consider graphical productions as being domain independent. We need to know of graphical compositions as things-in-themselves, which can be mapped to pictorial objects.

5.1.2. Representation of pictorial objects. Definition of the dimensionalities of pictures that can be read from graphical compositions, including geometries, topologies, and surface treatments. This definition has to be specific to particular dimensions but, for generality, it otherwise needs to be domain independent. Dimensions can be defined discretely, with their own demands on the graphic representation, and certain dimensions might be interrelated. More than one dimension might share some common instances of graphic entities – as occurs in the familiar phenomenon of multiple readings of depictions. Dimensions need to include calibrations so that values can be identified in graphical compositions. These syntactic parts of pictures form the content of a picture lexicon. Users' graphical expressions can then be mapped to lexical entries and values can be passed on to functional models for generating further semantic representations from pictures.

5.1.3. Representation of textual objects. Definition of character strings and their roles as 'parts of speech' (seen by users in compositions of text, and numerals), in terms of logical entities and their relations. The representation scheme has to accommodate logical differentiations for the presence of things (noun phrases), actions associated with things (verb phrases), identity (anaphoric reference I), plurals, tense, etc. (as possible syntactic categories of formal languages, including NL systems as computational sub-sets of human language). These syntactic parts of writing form the content of a word (or phrase) lexicon. Users' expressions can then be mapped to lexical entries and values can be passed on to functional models for generating further semantic representations from written compositions.

5.1.4. Functional models. Definition of tasks that can be applied to pictorial and textual objects, in terms of grammars or rule systems which correspond with generally accepted knowledge about treatments of those objects. These include systems for generating logical representations of NL semantics from users' expressions, systems for generating 3-D spatial projections from 2-D pictures, and system for establishing interrelationships between NL and spatial semantics. For generality, any functional models for establishing particular associations between pictorial and textual objects need to be valid for whole application domains. Results from these functional models need to be in a form that can be accommodated in a KB.

5.1.5. KB models. Definition of a general storage and retrieval environment which can accommodate values abstracted from users' expressions, via the lexica and processed through functional models – these are the values 'read' from the spatial differentiations exhibited in graphical productions. KB representations then are domain specific models resulting from a user's use of the system.

5.1.6. Artificial intelligence. The system's ability to execute mappings between an image representation, textual and pictorial objects, functional models, and KB models can be thought of as representing human intelligence – but the representation itself has to be viewed as artificial intelligence (with the implications noted earlier under role of computers). This ability rests on the system's in-built representation scheme (see later) which might exhibit and amplify behaviour which otherwise is thought to result from human cognitive processes – the system then serves as an instrument which amplifies certain mental powers of people.

The system's representation scheme has to maintain consistency and effect coherence across the above system components. This is a major research ambition, but a modest achievement is likely to suffice for the development of practical systems which allow users a lot of control over system behaviour.

5.2. System architecture

The above topics fit into a general system architecture as indicated in Fig. 7.

The main system components are a graphical production environment which behaves as a graphics machine for realising graphical expressions or screen images – these components, operating under the control of users, can be regarded as somewhat equivalent to people's normal practices in constructing images, whether or not they are using computers. People might use lexica for expressions in order to interpret them as words and pictures with associated grammar rules, to condition their further actions on expressions. Lexica support parsing of expressions to produce semantic representations in terms of further expressions that are acceptable to actions or functions that might be applied to a currently extant expression, or an image on the screen. Such

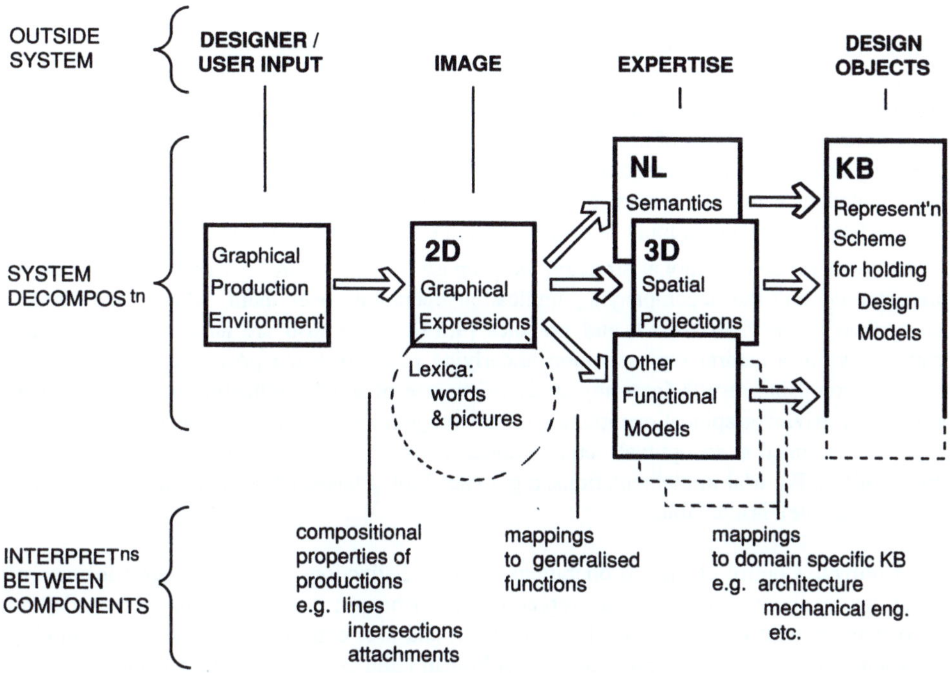

Fig. 7. A tentative general architecture for pointer systems

functions are represented by further system components, as functional models corresponding to human expertise on formal treatments of expressions. Results from functional models are deposited in the system's KB, as users' domain specific models which can be recalled as images on the screen. These latter system components offer computational processes for constructing and modifying users' expressions and, for this purpose, the behaviours of these components have to be known to users.

The architecture of a system has to include provision for communication between components, to map entities into the forms that can be used by the different components, and to maintain logical consistency and coherence across all representations held within the system. This ability of a system has to rest on a general representation scheme and its associated control features, which will be outlined in the next paragraphs.

5.2.1. Representation scheme. Construction of a knowledge representation scheme and the operations over knowledge representations needs to be guided by the general strategy for graphics outlined earlier. We want a representation scheme that can apply uniformly to all components of a system, covering graphic objects and derived pictorial and textual objects, plus functional models and KB models.

The representation scheme needs to include inferencing mechanisms of the kind now available in modern systems, such as those covering inheritance, defaults, updating, and constraint checking. These facilities must support relations between a graphical realisation of an image on a screen, and representations of depictions throughout the various components of the system, to its representations of users' domain knowledge. Relations also need to be supported in the reverse direction, to generate graphical realisations of words and pictures from existing states of representations within the system. Most critically, facilities of the representation scheme must accommodate the user's own input of expressions, in the form of words and pictures.

As a general computational strategy, we propose the use of DAG (directed acyclic graph) representations and the accompanying unification operations over them, which have already been applied in the fields of NL and graphics [9, 17]. DAGs also are an appropriate choice because they are a natural representation underlying object-oriented programming. Moreover, we expect promising results from any study of the use of graph unification as a fundamental operation underlying graphical manipulation, knowledge representation [1, 5], and NL processing [13], which is aimed at an optimal computational model of interaction between these different representations. For trial implementations, a graphics Prolog based representation scheme, MOLE (see later example), is available.

5.2.2. Control. Control of system operations rests on distinctions between: control features of formalisms (as built into a logical representation scheme); control features of modelling components (as built into prior functional models); and user control of models used to modify expressions. The first entails representations of human intelligence built into machines, and this can benefit from AI techniques. The second and third cases entail people's intelligent use of the former, with the implication that users have to know the behaviour of formalisms (and prior modelling components) which they employ on their expressions.

5.2.3. Control expressions. These are user actions aimed at operations which a user wishes to invoke within the system, with system defined interpretations: key hits, mouse clicks I on characters, special symbols, menus, and icons.

6. Examples

The following two examples are included to illustrate some limitations of current picture oriented interaction, and the potential of POINTER developments. One is taken from EdCAAD's experience on the ACORD project (mentioned earlier), and the other from early experience of integrated CAD systems.

The ACORD project has developed a prototype application representing transportation of goods, as shown in Fig. 8. Graphics depict towns (as nodes) linked by roads (as arcs). Graphic symbols denote towns, and they are labelled with charts showing quantities of goods at each town. Lorries are denoted by further symbols which can likewise be labelled with charts showing their contents. These lorries can be placed at towns and they can be placed on roads or moved to other towns. When they are at towns they can load or unload goods. NL and direct manipulation of graphics are used to say what is happening, and consequences appear in revised charts and changed positions of lorries. This application treats graphical entities as pre-defined primitives

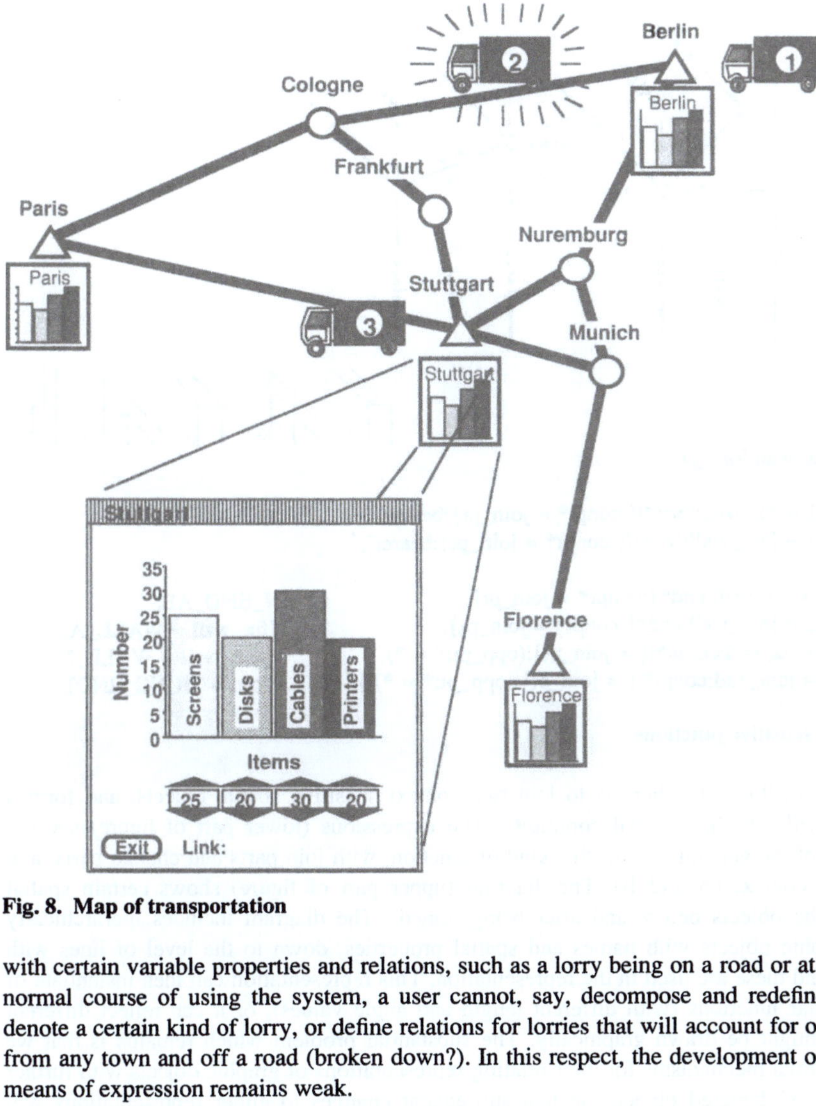

Fig. 8. Map of transportation

with certain variable properties and relations, such as a lorry being on a road or at a town. In the normal course of using the system, a user cannot, say, decompose and redefine a symbol to denote a certain kind of lorry, or define relations for lorries that will account for one being away from any town and off a road (broken down?). In this respect, the development of graphics as a means of expression remains weak.

Fig. 9 shows an example originally from CAD and reworked using MOLE (Modelling Objects with Logic Expressions [4, 8, 17]. This system allows users to define descriptions of objects and functions that can be applied to them, and includes graphic entities as possible objects. MOLE includes general treatments of inheritance and updating in a context of parts hierarchies and kind relations, such that any object can be a sub-part and a super-part, and a sub-kind (child) and a super-kind (parent) of other objects, and descriptions can include conditional and replacement expressions. A system component used to produce graphical images becomes a user of the MOLE representation scheme [15, 16].

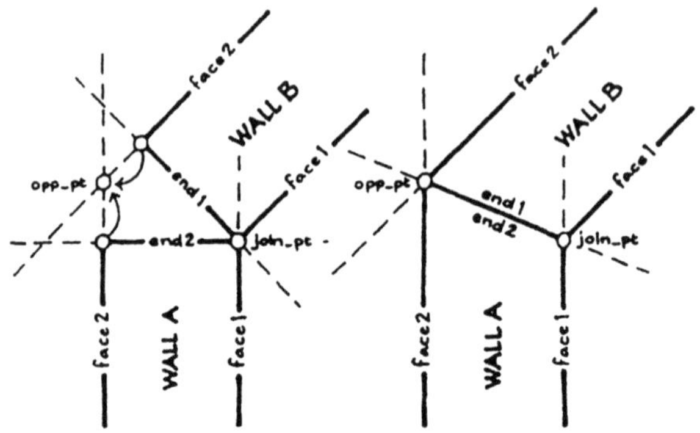

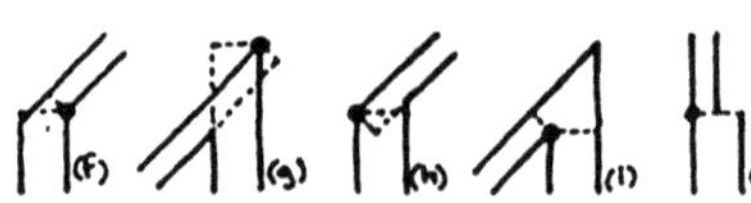


```
JUNC_END:
   [join_pt = fix_wall:join_pt,
    opp_pt:
       [conline1 = fix_wall:face*|(:conpt* = join_pt):bearer,
        conline2 = join_wall:face*|(:conpt* = join_pt):bearer],
    change:
       [fix_end = fix_wall:end*(:conpt* = join_pt),
        join_end = join_wall:end*(:conpt* = join_pt),
        move1 = fix_end:conpt*|( = join_pt):(opp_pt:* = *),
        move2 = join_end:conpt*|( = join_pt):(opp_pt:* = *)]].
```

```
JOIN_END_AB:
   [fix_wall = WALL_A,
    join_wall = WALL_B,
    join = ~JUNC_END].
```

Fig. 9. Context sensitive junctions

The problem illustrated here is to join two context sensitive spatial objects, and form a junction that will establish spatial continuity. The expressions (lower part of figure) show a representation of the general case of this kind of junction, with join parts and change parts, and an instance in context (A and B). The diagram (upper part of figure) shows certain spatial properties of the objects before and after being joined. The diagram includes hierarchically structured graphic objects with names and spatial properties, down to the level of lines with angle values, and these are used in the representation. This representation can then instantiate or modify particular junctions (with different length and angle values), or it can reflect different instances that might be drawn graphically. The substantial problem which remains is that we lack orderly formal mechanisms for inter-relating representations of graphic objects with further representations of depicted objects, so that subsequent changes to either representations will have effects that can be known and anticipated by users.

7. Conclusion

This paper has outlined a major programme for HCI research. Its starting point rests on a distinction between human knowledge, within persons, and expressions as representations of that knowledge, external to persons. Humans use expressions to get at and share each other's knowledge, and we can do so because we are able to appreciate the distinction between overt expressions and our sense of knowing within each of us – a distinction that is evidently employed by designers, and managers and policy makers.

Computers do not sense this distinction. They are machines for performing operations on expressions and we can use them as tools for constructing and modifying our expressions. Computers remain just such machines, even when their operations on many expressions contributed by different persons make them look like they are behaving intelligently. The role of AI techniques has been identified with a system's inferencing mechanisms associated with its general representation scheme, within the system, which condition its surface behaviour as evident to users. These facilities have to maintain logical consistency and coherence across the system's representations of user expressions, and they condition system responses to user actions.

The focus for research on HCI then has to be on the possible representations within computers, of user expressions, so that a system's behaviour can be intelligible and controllable by users. This focus is central to all potentially practical computer systems, and it provides scope for a major programme of collaborative research projects.

References

1. H. Ait-Kaci and R. Nasr, "LOGIN: A Logic Programming Language with Built-in Inheritance," *Journal of Logic Programming*, vol. 3, 1986.
2. A. Bijl, D. Stone, and D.H. Rosenthal, "Integrated CAAD Systems," EdCAAD Report for the Dept. of the Environment, University of Edinburgh, 1979.
3. A. Bijl and L.A. Pineda, "Notions of Language and Design for Intelligent CAD," in *Intelligent CAD II*, ed. T. Tomiyama and T. Holden, North-Holland, 1989.
4. A. Bijl, *Computer Discipline and Design Practice — Shaping Our Future,* Edinburgh University Press, 1989.
5. J. Fargues, A.M.C. Landon, A. Dugourd, and L. Catach, "Conceptual Graphs for Semantics and Knowledge Processing," *IBM Journal for Research and Development*, vol. 30, 1986.
6. E.M. Hoskins, "The OXSYS System," in *Computer Applications in Architecture, Applied Science*, ed. J.S. Gero, pp. 343-391, 1977.
7. E. Klein, "Dialogues with Language, Graphics and Logic," in *ESPRIT '87: Achievements and Impact*, North-Holland, 1987.
8. R. Krishnamurti, "The MOLE Picture Book: On a Logic for Design," *Design Computing*, vol. 1, no. 3, 1986.
9. J.R. Lee, A. Bijl, and P.J. Szalapaj, "The Graphics Component of the ACORD System," EdCAAD/ACORD working paper, 1986.
10. J. Opas and M. Mantyla, "Introducing Manufacturing Knowledge into Intelligent CAD Systems," *2nd IFIP WG 5.2 Workshop on Intelligent CAD*, Cambridge, 1988.
11. L.A. Pineda, E. Klein, and J.R. Lee, "GRAFLOG: Understanding Drawings Through Natural Language," *Computer Graphics Forum*, vol. 7, 1988.

12. A.A.G. Requicha, ''Geometric Modelling and Programmable Automation,'' *IFIP TC5 conference on CAD/CAM Technology Transfer to Latin America*, Mexico City, 1988.

13. S.M. Shieber, ''An Introduction to Unification-based Grammar Formalisms,'' *CSLI Lecture Note Series*, 1986.

14. P. van Sommers, *Drawing and Cognition,* Cambridge University Press, 1984.

15. P. Szalapaj, ''A Transformational Grammar for Line Drawings,'' *int. symp. on Fuzzy Systems and Knowledge Engineering*, Guangzhou/Guiyang, China, 1987.

16. P. Szalapaj, J.R. Lee, and A. Bijl, ''The Semantics of Computational Systems that Depict Non-Graphical Information,'' in *Esprit P.393 ACORD deliverable T3.6b*, 1988.

17. C. Tweed and A. Bijl, ''MOLE: A Reasonable Logic for Design?,'' in *Intelligent CAD Systems II – Implementational Issues*, ed. V. Akman, P.J.W. ten Hagen, and P.J. Veerkamp, Springer-Verlag, 1989.

18. F. Zdybel, N.R. Greenfeld, M.D. Yonke, and J. Gibbons, ''An Advanced Information Presentation System,'' *Computer Graphics 81*, pp. 19-36, 1981.

17

Intelligent User Interface for Intelligent CAD

Paul ten Hagen, and Zsófia Ruttkay[†]

Department of Interactive Systems
Centre for Mathematics and Computer Science (CWI)
P.O. Box 4079, 1009 AB Amsterdam, The Netherlands

Abstract: A user cannot fully exploit the capabilities of an intelligent CAD system unless the user interface is equally powerful. In this paper we list the general and design-related criteria of intelligent communication. The emphasis is on the semantics of the inputs/outputs in context of communication, the design process, and the design constraints. We give a proposal how to extent a traditional user interface with a monitor component in order to meet the requirements of intelligent communication. The proposal is an implementational model for the communicational tasks of intelligent CAD.
Keywords: Human Computer Interaction, Intelligent Communication, Direct Communication.

---

1. Introduction

1.1. User interface issues of intelligent CAD systems

Attempts to improve CAD systems by applying techniques from the domain of artificial intelligence have addressed almost every aspect of CAD. This paper is concerned with ways to make the user of an interactive CAD system aware of such improvements. It is obvious that increased capabilities must become accessible through the user interface in order to be useful. Providing access to the various new CAD system functions is far from trivial, in fact it poses several so far unsolved problems.

In order to describe the size of the problem we will describe some of the most important issues in intelligent CAD (ICAD) and the corresponding user interface issues that follow from them. An ICAD system should be knowledgeable in several ways [1, 7]:

[†] Computer and Automation Institute, Hungarian Academy of Sciences, P.O.B. 63, H-1502 Budapest, Hungary

- It maintains different models of the object being designed, such as functional, structural and geometric models.
- It represents the evolution of the design, the partial designs and alternatives in the various design stages.
- It handles domain specific knowledge for instance about constraints among design elements.

Logic and object oriented programming methods provide a more powerful symbolic computational aid to provide such a variety of abstract views on a model [2]. They also can provide better means to represent the relations between all these variants of the design object. Moreover, this can be done in one integrated system.

Nevertheless, these systems still require the user to be in control. The CAD system is a (slightly) better assistant, provided that the user can easily express his commands. An important function of the system with this more complex representation of the design object, is to automatically select relevant information. There is always a limited capacity to visualise information, hence no resources are to be wasted. Moreover, showing relevant information is a sure sign that the system understands the user.

These characteristics of ICAD raise novel user interface issues. They all are concerned with presenting (more) semantic information in addition to illustrating the design object. This immediately results in a richer dialogue language, because everything that can be presented can also be discussed. Below we will list a number of these issues. Their relation with the improved functionalities given above is quite obvious.

Concerning the topic of conversation:

- Means must be provided to define and present relations in and between models.
- Data items must be annotated with qualifiers such as age, importance and precision. The user interface can be made responsible for providing some of these automatically.
- The interface must be able to handle several design object presentations, either simultaneously or sequentially.

Concerning the flow of the dialogue:

- Embedded sub-dialogues and multiple threads in dialogues must be supported. For instance, the system can enter an extra dialogue for maintaining consistency.
- An ICAD system is used by expert users. They must be allowed to tune the system to their way of working. An untuned system is (by definition) what a non-expert user gets.
- The system must be capable of visualising the consequences of user inputs, including the reasons, e.g., domain specific reasons.
- The user interface must be able to record history, in particular to explain when and why decisions were made.

Concerning the language of communication:

- The mapping between CAD operations and user language may change as the dialogue (and the design) evolves. The mapping can be many to many [6].
- The vocabulary and presentation methods may also change.

In this paper we will investigate how these requirements can be met by primarily looking at the user interface itself. This is of course not the only aspect to be considered. However, the result of our analysis might be that if the CAD system is capable of improved functionality, than the user interface can convey it.

1.2. A separate user interface component

Application programs which are heavily dependent on user interaction, nowadays have a separate component for the user interface [5]. Such a component can be based on a specialized system, called the user interface management system. It embodies a method to abandon poorly designed ad hoc user interfaces. It provides a rich set of basic interaction techniques, including window management functions.

A user interface management system (and the separate interface module) provide a number of advantages which will be heavily exploited to get an intelligent interface, e.g. an interface which can also use AI techniques.

- A clear model of interaction. All user interface functions are present in one environment. Hence, all combinations are in that environment as well. Thus a selective set can be provided for a given task, providing consistency in language across tasks, etc.
- Device independence. Every basic interaction technique can be chosen according to the given hardware. Above this basic level all device specific properties are shielded away.
- Interaction style. To some extend the interactions can be shaped according to individual user (group) requirements. For instance each can make its own screen lay out arrangements.

In the case of ICAD systems there is one major capability to be added to the user interface. It is the ability to access and process semantic information from the application directly in the user interface, for no other purpose than to moderate the interactions themselves. Thus, syntax and semantics get coupled visibly and more directly.

This capability will form the basis for many improvements in the interface itself. It constitutes a bridge between the application and the user interface. The reason for this bridge is twofold:

- In order to decide what is to be displayed, both the application status and the user inputs have to be considered simultaneously.
- Maximum directness of interaction also requires simultaneous interpretation of both user input and application status.

In the next sections we will first outline what kind of functionality can be improved by or provided by this capability, either from the point of view of general user interface properties or from the point of view of the ICAD functionality. It is up to the builders of ICAD systems to achieve the goals of more intelligence in the CAD interface exploiting the bridge function. This is similar to the view how to use a user interface manager. It provides means for building good interfaces, but guarantees nothing for any particular interface built with it.

After this overview an extension of the user interface manager will be introduced. For the sake of clarity all new functionality associated with the bridge function will be provided in yet another separate module called the user interface monitor.

2. Interface Functions

We will introduce the monitor functionality using user interface function examples. We will start with a very simple example to give the general idea.

Suppose a designer has to place a table in a room. Given the situation in the room he may change the table shape. We only consider the table top. Relevant information:

- number of persons to be seated;
- table length;
- table width.

The user is allowed to deviate from a rectangular shape only if all rectangular solutions would fail.

table = n–of–persons ∧ length ∧ width

The transaction cell fragment states that for defining a table three units of information are necessary. Each of these units is delivered by the corresponding process. The ∧-operator indicates that these units are accepted in any order.

The efficiency and correctness of the interaction now depends on the behaviour of the three separate processes n–of–persons, length and width. The monitor as envisaged by us can contribute considerably to the effectiveness of these processes as well as to their combined effect denoted by the transaction process table.

2.1. Example: default values

Assume the user is operating the input mechanisms for length and width. The application has a procedure to determine from these the number of persons that can sit at the table. By passing this result to the n–of–persons process, this process can propose the result as default value. As a result the user interaction with this process is reduced to a confirmation. Of course the user may decide to override the default and request a different capacity for the table.

In a similar fashion the user might have decided about the number of persons first. This would then lead to defaults for length and width. Below we give the transactions necessary to have the monitor communicate these results among the various processes involved.

U–tabledraw ⇒ (length, width)

The user by manipulating the drawing facilities produces information for the processes length and width. Similarly he can produce a numerical value for the process n–of–persons:

U–number ⇒ n–of–persons

As we have seen in the first rule, the table process makes a 'complete table' from the values produced by the right-hand sides of the input processes.

Now we introduce two default producing configurations:

(length ∨ width) ⇒ Mon–n–of–p ⇒ n–of–persons

and

n–of–persons ⇒ Mon–landw ⇒ (length ∧ width).

The process Mon–n–of–persons resides in the monitor and tries to deduce a value for number of persons from the table shape. If it succeeds this value is passed on to the appropriate process. Similarly the monitor process landw tries to adapt the shape of the table to an explicitly given value for number of persons. The values thus provided by the monitor are marked as default values. Only after confirmation by the user they are promoted to normal value status.

The connections indicated by ⇒ can be identified in such a way that the receiving process can differentiate the input source. In this way the distinction for instance between user input and system generated input can be maintained. The hard part of the work to be done for providing default facilities is in providing the right control structure. The kernel for these are in the transaction rules. This will remain so for other features of intelligent interfaces.

2.2. Example: reuse of values

Assume first the rule

U–tabledraw ⇒ (length, width)

has been activated. This in turn may have caused activation of

(length ∨ width) ⇒ Mon–n–of–p ⇒ n–of–persons.

As a result we now have also a proposed value for n–of–persons. Therefore the rule

table = n–of–persons ∧ length ∧ width

fires, and the table process can produce a view of the current table. Now the user operates the number input for persons:

U–number ⇒ n–of–persons,

thus overriding the default value.

The monitor will observe that the new figure for persons is inconsistent with the table shape. And it will propose new default values for length and width. However in doing so it will reuse the existing values in such a way that the new shape is similar to the previous one. The monitor has embodied the notion of similarity in the rules for reuse of shape values. For instance it may be part of a transformation procedure for tables which changes their size but not the shape. All this will be the outcome of another activation of the rule

n–of–persons ⇒ Mon–landw ⇒ (length ∧ width).

If we now reconsider the original rule:

table = n–of–persons ∧ length ∧ width,

then, by adding the two monitor processing rules we have created the options for the user to have a system which adapts itself by generating default values. The user can either start of with a given shape or with a given size and can switch between these two modes any time.

3. The Formal Model

3.1. The service user interface

The service user interface is based on the Dialogue Cell system [3,4]. This is user interface management system which allows for three-dimensional graphics objects to appear as tokens in dialogues. The basic notions of the service user interface are the units of communication acts described by so-called *Transaction Cells.* A transaction cell reads and writes symbols while active and ultimately generates its result symbol.

The symbols communicated by a transaction cell are exchanged between named processes after having established a communication with them. A special set of these processes are the slave transaction cells which have been activated by this (parent) cell. Slave cells are deactivated when their parent is deactivated. Some of the descendant cells are basic input and output processes to directly communicate with the user.

A transaction cell can be characterized as a three component unit:

- transaction cell name, and symbol type;
- activation and connection rule;
- read/write grammar rule.

The transaction cell name is an external name to be used by all other units (including the parent) who wish to communicate. The activation rule specifies all descendant transactions that need to be activated, as well as already active processes with which a communication link must be established. It also precisely describes when during the active life time of the transaction cell these links must exist. Note that this constitutes the basic mechanism to specify when a user is allowed to perform what action. The activation rule allows for simultaneous as well as sequential connections.

The read/write grammar rule is a regular expression specifying in what order reads and writes may happen. It allows for sequential as well as parallel reads and writes. The specification of the reads and writes is entirely symbolic. The actual values being communicated follow automatically from the real time values produced by the various processes including the basic input/output processes.

A mapping from, say, input to an internal representation, which is communicated further, is also specified through a write-read sequence to/from the mapping function-process. Equally the visualisations occurring during interactions are taken care of by graphics processes connected to a transaction cell.

In this way we have a model for the user interface service component which abstracts from all process details as long as a number of abstract communication functions exist, being: *read, write, activate, deactivate, connect* and *disconnect.* All real activities are carried out inside processes. Transaction cells allow us to only but completely control how these processes communicate.

The service user interface executes all processes concerned directly with outside (e.g. user-) communication. However, these processes can, without restriction, be engaged in transactions with processes outside the service user interface.

3.2. The monitor

The monitor can be seen as a collection of additional processes which contain chunks of application information, and chunks of user interface information. In particular they maintain:

- values of user and application variables with quantifiers such as: *default, user supplied, suggested, confirmed,* etc.
- value histories; for instance for reuse.
- context descriptions, e.g. which of the monitor processes are currently active, (inactive monitor processes do not receive any data or events).

The monitor also contains processes to initiate visualization by the user interface of state variables, including relations.

4. Semantic Processing in the User Interface

We will now extend the monitor with semantic processing capabilities and show through the extended example of §2, how these processes can be connected to the user interface processes of the service and monitor module. These new processes will also run under the regime of the monitor.

Assume that each time a table has been specified, the user can position the table in a room. This is because the table function is part of a workplace design system. The room-process accepts positions of objects such as tables. Next an evaluation function exists which can analyse how well the table would fit the room. There are other objects (e.g an archive) which can be added to the room independently from table. The evaluation function can analyse these in a similar fashion.

Each time the table is placed the following rule fires:

(table $\wedge$ room) $\Rightarrow$ evaluate.

The result of evaluate is a rating of the placement of the table with respect to its position relative to the room and the other objects in it. Poor results (bad placement) are reported relative to the table. In case of the rule

(archive $\wedge$ room) $\Rightarrow$ evaluate,

the same evaluation will be carried out, but the result will indicate why the archive is not well-placed. Hence the system has a notion of the object being considered.

This notion is very hard to support in user interfaces and yet is a very basic one for understanding user intentions. In the process structuring we have provided, each object can become the considered one by activating the appropriate process configuration. Hence, we have a representation for it. The next question is whether we can generate the appropriate configuration at all times.

The default object is the most recent one that was altered. Let us assume that there is a user device for positioning objects in the room, then we have the following generic rules:

$$\text{U–pos.X} \Rightarrow \text{room}$$
$$(\text{X} \wedge \text{room}) \Rightarrow \text{evaluate},$$

where X can be table, archive, etc.

Now the system can cluster these rules such that a reposition of an object will do evaluate relative to it. The natural cause of events may be that positioning of the table is part of the table manipulations next to length, width and n–of–persons. Hence there may also be a monitor function for recalculating the table position after reshaping.

$$(\text{length} \vee \text{width}) \Rightarrow \text{Mon–pos.table} \Rightarrow \text{room}.$$

Recall that changing number of persons may also cause the table to be reshaped. An chain of derived actions may thus end up in a new evaluation of the room.

4.1. Example: fixed object

The user can explicitly request that the considered object is the archive even when the table is being manipulated. This would override the above cluster with:

$$\text{U–pos.table} \Rightarrow \text{room}$$
$$(\text{archive} \wedge \text{room}) \Rightarrow \text{evaluate}.$$

Hence, manipulating any object would still end in the room being evaluated with respect to archive.

4.2. Example: direct manipulation

We now can formulate direct manipulation configurations:

$$\text{U–table} \Rightarrow (\text{table} \wedge \text{room}) \Rightarrow \text{evaluate}.$$

Where:

$$\text{U–table} = (\ \text{U–tabledraw} \Rightarrow (\text{length, width}) \vee$$
$$\text{U–number} \Rightarrow \text{n–of–persons} \vee$$
$$\text{U–pos.table}$$
$$)$$

Any manipulation concerning the table will immediately lead to an evaluation which can be instantaneously visualised. Recall that the monitor functions behind the scene are essential contributors as well because they will produce a new table position from, for instance, length. Direct manipulation can easily be combined with fixed focus on objects, so that one can shift, e.g. the archive and instantaneously see a reevaluation of the table in the room.

5. Conclusion

We have indicated that a good strategy for building user interfaces is by providing a mechanism for generic rule based control structuring of essentially the same processes. Such a mechanism can be provided starting from Transaction Cells. This suggests that it is a feasible approach.

References

1. V. Akman, P.J.W. ten Hagen, and P.J. Veerkamp (eds.), *Intelligent CAD Systems II, Implementational Issues,* Springer-Verlag, Berlin, 1989.
2. P.J.W. ten Hagen and T. Tomiyama (eds.), *Intelligent CAD Systems I, Theoretical and Methodological Aspects,* Springer-Verlag, Berlin, 1987.
3. P.J.W. ten Hagen and R. van Liere, ''A Model for Graphical Interaction,'' CWI-report CS-R8718, Amsterdam, 1987.
4. R. van Liere and P.J.W. ten Hagen, ''Introduction to Dialogue Cells,'' CWI-report CS-R8703, Amsterdam, 1987.
5. G.E. Pfaf (ed.), *User Interface Management Systems,* Springer-Verlag, Berlin, 1985.
6. Zs. Ruttkay, ''Multi-media Presentation in CAD Systems,'' in *Intelligent CAD Systems II, Implementational Issues*, ed. V. Akman P.J.W. ten Hagen, and P.J. Veerkamp, pp. 77-92, Springer-Verlag, Berlin, 1989.
7. H. Yoshikawa and D. Gossard (eds.), *Intelligent CAD, I,* North-Holland, Amsterdam, 1989.

18

Discussion Session: User Interface

Paul Veerkamp

1. Papers

The closing session is represented by two papers, under the heading of User Interface. Both papers are concerned with how a designer can express his intentions to an intelligent CAD system. The former by Aart Bijl advocated an instrumentalist approach to computer systems. The latter by Paul ten Hagen focussed on a dialogue language for human computer interaction. Bijl's presentation particularly stressed the importance for a designer to express his thoughts both by words and pictures. A designer must, therefore, be given a tool which allows him to construct expressions which give him control over the behaviour of the system.

Discussion of this presentation then questioned whether or not such a system is considered 'intelligent' when the only task of the system is to execute the commands given by the designer. Bijl stressed that, in his opinion, a CAD system may never limit the designer's abilities. The intelligence of the system lies therefore in understanding the designers expressions and acting adequately upon his commands and *not* in performing certain design tasks autonomously.

Ten Hagen's approach took a slightly different position. He argued that a system is entitled to perform certain tasks autonomously provided that the designer is always informed about what the system is doing. A monitor component, having both a model of the user and the system, guides the communication between them. The monitor knows how to act upon the designers demands since it knows the status of the system and the user's inputs. Transaction cells are a formal means to describe the communication.

Arbab opened the discussion on this paper by challenging the need for yet another language for ICAD. Since we have ADDL, etc. for representing knowledge for ICAD systems, what is ICAD really?[1] What are the requirements that hold for ICAD languages? Ten Hagen replied that

[1] Editor's note: Paul ten Hagen is from the same group that developed ADDL.

developing a new formalism is a good exercise to grasp the essential requirements for a communication language. As soon as these requirements are made explicit, they can be integrated in an ICAD language like ADDL.

MacCallum then commented that the definition of consistency and completeness must be made in the design context and *not* in the far more rigid mathematical/logic context. Ten Hagen gave an example of constraint manipulation in a geometric reasoning system. The intelligence of the system can be measured by the way it treats problems that it cannot solve. In that case the system must ask the designer and explain what the problem is, so that the designer can try to solve the problem himself.

2. General Discussion

The importance of an 'intelligent' user interface for ICAD was discussed. An issue which was raised was the control of the flow of information. Does the user interface component dictate the system or vice versa? The standpoint was taken that neither the user interface nor the system is in command, the designer directs the design process. Both the user interface and the ICAD system have a task for which it is responsible.

Barthès raised the issue that user interfaces must be regarded as filters which should reduce the amount of input of the designer intelligently. He considered a user interface intelligent when it adapts to the designer's behaviour by applying common sense reasoning. Another criterion for intelligent behaviour is that the user interface is able to accept erroneous or incomplete input, i.e. it doesn't halt in such a case but it tries to recover.

Arbab concluded that the user interface is the system's component which really acts as an intelligent aid to the designer. It visualises the system's behaviour. When the ICAD system recognises that it can no longer solve the design problem, it is the user interface's task to ask for guidance. The problem is then clarified in dialogue with the designer until the ICAD system can proceed with the problem.

List of Authors

Patrice Anota
Université de Technologie de Compiègne, Dépt. de Génie Informatique, C.N.R.S. UA 817, BP 233, 60206 Compiègne, France.

Farhad Arbab
Computer Science Department, SAL 200, MC 0782, University of Southern California, Los Angeles, CA 90089, USA.

Keith Baker
Dept. of Computer Science, University of Reading, Reading, Berkshire, RG6 2AX, UK.

Linden Ball
Department of Computing, Plymouth Polytechnic, Plymouth PL4 8AA, UK.

Jean-Paul A. Barthès
Université de Technologie de Compiègne, Dépt. de Génie Informatique, C.N.R.S. UA 817, BP 233, 60206 Compiègne, France.

Can A. Baykan
Center for Integrated Manufacturing Decision Systems, The Robotics Institute, Carnegie Mellon University, Pittsburgh, PA. 15213, USA.

Aart Bijl
EdCAAD, Univ. of Edinburgh, 20 Chambers Street, Edinburgh EH1 1JZ, UK.

Phil Culverhouse
Department of Computing, Plymouth Polytechnic, Plymouth PL4 8AA, UK.

Ian Dennis
Department of Computing, Plymouth Polytechnic, Plymouth PL4 8AA, UK.

Rajiv S. Desai
Artificial Intelligence Group, Jet Propulsion Laboratory, California Institute of Technology, Pasadena, CA 91109, USA.

Rajkumar S. Doshi
Artificial Intelligence Group, Jet Propulsion Laboratory, California Institute of Technology, Pasadena, CA 91109, USA.

Kamal El Dahshan
Université de Technologie de Compiègne, Dépt. de Génie Informatique, C.N.R.S. UA 817, BP 233, 60206 Compiègne, France.

Jonathan Evans
Department of Computing, Plymouth Polytechnic, Plymouth PL4 8AA, UK.

Mark S. Fox
Center for Integrated Manufacturing Decision Systems, The Robotics Institute, Carnegie Mellon University, Pittsburgh, PA. 15213, USA.

Yoshiki Ishida
Department of Precision Machinery Engineering, Faculty of Engineering, The University of Tokyo, Hongo 7-3-1, Bunkyo-ku, Tokyo 113, Japan.

Peter Jagodzinski
Department of Computing, Plymouth Polytechnic, Plymouth PL4 8AA, UK.

Ewan H. Klein
EdCAAD and Centre for Cognitive Science, University of Edinburgh, 20 Chambers Street, Edinburgh EH1 1JZ, UK.

Raymond K. Lam
Artificial Intelligence Group, Jet Propulsion Laboratory, California Institute of Technology, Pasadena, CA 91109, USA.

Ken J. MacCallum
University of Strathclyde, CAD Centre, 131 Rottenrow, Glasgow G4 0NG, UK.

Pat Pearce
Department of Computing, Plymouth Polytechnic, Plymouth PL4 8AA, UK.

Ravic Pieters Kwiers
Department of Interactive Systems, Centre for Mathematics and Computer Science (CWI), Kruislaan 413, 1098 SJ Amsterdam, The Netherlands.

Luis A. Pineda
EdCAAD and Centre for Cognitive Science, University of Edinburgh, 20 Chambers Street, Edinburgh EH1 1JZ, UK.

Jan Rogier
Department of Interactive Systems, Centre for Mathematics and Computer Science (CWI), Kruislaan 413, 1098 SJ Amsterdam, The Netherlands.

Zsófia Ruttkay
Computer and Automation Institute, Hungarian Academy of Sciences, P.O.B. 63, H-1502 Budapest, Hungary.

Dean Scothern
Department of Computing, Plymouth Polytechnic, Plymouth PL4 8AA, UK.

Pär Silén
Technical Research Centre of Finland, Laboratory of Urban Planning and Building Design, Itätuulenkuja 11, SF-02100 Espoo, Finland.

Tapio Takala
Helsinki University of Technology, Department of Computer Science, SF-02150 Espoo, Finland.

Paul ten Hagen
Department of Interactive Systems, Centre for Mathematics and Computer Science (CWI), Kruislaan 413, 1098 SJ Amsterdam, The Netherlands.

Tetsuo Tomiyama
Department of Precision Machinery Engineering, Faculty of Engineering, The University of Tokyo, Hongo 7-3-1, Bunkyo-ku, Tokyo 113, Japan.

Jan Treur
Expert Systems Section, Programming Research Group, University of Amsterdam, Kruislaan 409, 1089 SJ Amsterdam, The Netherlands.

Paul Veerkamp
Department of Interactive Systems, Centre for Mathematics and Computer Science (CWI), Kruislaan 413, 1098 SJ Amsterdam, The Netherlands.

Gill Venner
Department of Computing, Plymouth Polytechnic, Plymouth PL4 8AA, UK.

Bin Wang
Computer Science Department, SAL 200, MC 0782, University of Southern California, Los Angeles, CA 90089, USA.

Deyi Xue
Department of Precision Machinery Engineering, Faculty of Engineering, The University of Tokyo, Hongo 7-3-1, Bunkyo-ku, Tokyo 113, Japan.

EurographicSeminars

Tutorials and Perspectives in Computer Graphics

Eurographics Tutorials '83. Edited by P. J. W. ten Hagen.
XI, 425 pages, 164 figs., 1984

User Interface Management Systems. Edited by G. E. Pfaff.
XII, 224 pages, 65 figs., 1985

Methodology of Window Management. Edited by F. R. A. Hopgood, D. A. Duce, E. V. C. Fielding, K. Robinson, A. S. Williams.
XV, 250 pages, 41 figs., 1985

Data Structures for Raster Graphics. Edited by L. R. A. Kessener, F. J. Peters, M. L.P. van Lierop. VII, 201 pages, 80 figs., 1986

Advances in Computer Graphics I. Edited by G. Enderle, M. Grave, F. Lillehagen.
XII, 512 pages, 168 figs., 1986

Advances in Computer Graphics II. Edited by F. R. A. Hopgood, R. J. Hubbold, D. A. Duce. X, 186 pages, 96 figs., 1986

Advances in Computer Graphics Hardware I. Edited by W. Straßer.
X, 147 pages, 76 figs., 1987

GKS Theory and Practice. Edited by P. R. Bono, I. Herman.
X, 316 pages, 92 figs., 1987

Intelligent CAD Systems I. Theoretical and Methodological Aspects.
Edited by P. J. W. ten Hagen, T. Tomiyama. XIV, 360 pages, 119 figs., 1987

Advances in Computer Graphics III. Edited by M. M. de Ruiter.
IX, 323 pages, 247 figs., 1988

Advances in Computer Graphics Hardware II. Edited by A. A. M. Kuijk, W. Straßer.
VIII, 258 pages, 99 figs., 1988

CGM in the Real World. Edited by A. M. Mumford, M. W. Skall.
VIII, 288 pages, 23 figs., 1988

Intelligent CAD Systems II. Implementational Issues. Edited by V. Akman, P. J. W. ten Hagen, P. J. Veerkamp. X, 324 pages, 114 figs., 1989

Advances in Computer Graphics IV. Edited by W. T. Hewitt, M. Grave, M. Roch. XVI, 248 pages, 138 figs., 1991

Advances in Computer Graphics V. Edited by W. Purgathofer, J. Schönhut. VIII, 223 pages, 101 figs., 1989

User Interface Management and Design. Edited by D. A. Duce, M. R. Gomes, F. R. A. Hopgood, J. R. Lee. VIII, 324 pages, 117 figs., 1991

Advances in Computer Graphics Hardware III. Edited by A. A. M. Kuijk. VIII, 214 pages, 88 figs., 1991

Advances in Object-Oriented Graphics I. Edited by E. H. Blake, P. Wisskirchen. X, 218 pages, 74 figs., 1991

Advances in Computer Graphics Hardware IV. Edited by R. L. Grimsdale, W. Straßer. VIII, 276 pages, 124 figs., 1991

Advances in Computer Graphics VI. Edited by G. Garcia, I. Herman. IX, 449 pages, 186 figs., 1991

Intelligent CAD Systems III. Practical Experience and Evaluation. Edited by P. J. W. ten Hagen, P. J. Veerkamp. X, 270 pages, 116 figs., 1991

MIX
Papier aus verantwortungsvollen Quellen
Paper from responsible sources
FSC® C105338

If you have any concerns about our products,
you can contact us on
ProductSafety@springernature.com

In case Publisher is established outside the EU,
the EU authorized representative is:
Springer Nature Customer Service Center GmbH
Europaplatz 3, 69115 Heidelberg, Germany

Printed by Libri Plureos GmbH
in Hamburg, Germany